SPACE: THE FIRST QUARTER CENTURY

In the first quarter century of space flight, thousands upon thousands of craft were propelled beyond the earth's atmosphere to perform a vast variety of tasks, limited, it sometimes seemed, only by human imagination. This chronology highlights the most thrilling of these endeavors: manned space flight, the conquest of the moon and the programs of deep-space probes into the universe beyond. Passages printed in blue denote a first achievement.

1957

OCT. 4 ● SPUTNIK I (USSR) First man-made earth satellite.

NOV. 3 ● SPUTNIK II (USSR) First satellite to collect biological data from orbit; carried dog Laika.

1958

JAN. 31 ● EXPLORER I (USA) First American earth satellite.

MAR. 17 ● VANGUARD (USA) First satellite to use solar power.

MAY 15 ● SPUTNIK III (USSR) First comprehensive geophysical data from orbit.

1959

JAN. 2 ● LUNA 1 (USSR) First spacecraft to achieve earth-escape velocity; missed the moon and went into orbit around the sun.

MAR. 3 ● PIONEER 4 (USA) First deep-space probe; passed within 37,300 miles of the moon.

MAY 28 ● JUPITER (USA) First primates in space (Able and Baker); suborbital.

SEPT. 12 ● LUNA 2 (USSR) First lunar probe to impact on moon; no data returned.

OCT. 4 ● LUNA 3 (USSR) First lunar probe to photograph moon's far side.

1960

AUG. 19 ● SPUTNIK V (USSR) First recovery of orbited animals from space (dogs Strelka and Belka).

1961

JAN. 31 ● MERCURY-REDSTONE 2 (USA) First test of Mercury-Redstone flight vehicle with passenger aboard (chimpanzee Ham); suborbital.

FEB. 12 ● VENERA 1 (USSR) Venus probe; passed within 62,000 miles of Venus.

APR. 12 ● VOSTOK 1 (USSR) First manned space flight; Yuri Gagarin; one orbit; 1 hour 48 minutes.

MAY 5 ● MERCURY-REDSTONE 3 (USA) First American manned suborbital flight; Alan Shepard; 15 minutes 22 seconds.

JULY 21 ● MERCURY-REDSTONE 4 (USA) Manned suborbital flight; Gus Grissom; 15 minutes 37 seconds.

AUG. 6 ● VOSTOK 2 (USSR) Manned space flight; Gherman Titov; 16 orbits; 25 hours.

1962

FEB. 20 ● MERCURY-ATLAS 6 (USA) First American manned orbital flight; John Glenn;

three orbits; 4 hours 54 minutes.

APR. 23 ● RANGER 4 (USA) First American lunar probe to impact on moon; equipment failed; no pictures returned.

MAY 24 ● MERCURY-ATLAS 7 (USA) Manned mission; Scott Carpenter; three orbits; 4 hours 54 minutes.

AUG. 11 ● VOSTOK 3 (USSR) Part of first Soviet dual mission (with Vostok 4); A. Nikolayev; 64 orbits; 3 days 22 hours.

AUG. 12 ● VOSTOK 4 (USSR) Part of first Soviet dual mission (with Vostok 3); came within 3.1 miles of Vostok 3 on first orbit; P. Popovich; 48 orbits; 2 days 23 hours.

AUG. 27 ● MARINER 2 (USA) First successful fly-by of Venus.

OCT. 3 ● MERCURY-ATLAS 8 (USA) Manned mission; Walter Schirra; orbits increased to six; 9 hours.

1963

MAY 15 ● MERCURY-ATLAS 9 (USA) First American manned flight to exceed 24 hours; Gordon Cooper; 22 orbits; 34 hours 20 minutes.

JUNE 14 ● VOSTOK 5 (USSR) Part of dual mission (with Vostok 6); V. Bykovsky; 81 orbits; 5 days 23 hours.

JUNE 16 ● VOSTOK 6 (USSR) Dual mission (with Vostok 5); came within 3 miles of Vostok 5; Valentina Tereshkova (first woman in space); 48 orbits; 2 days 22 hours 48 minutes.

1964

JULY 28 ● RANGER 7 (USA) First successful American lunar probe; impacted on moon; returned 4,316 close-up photos of lunar surface down to impact.

OCT. 12 ● VOSKHOD I (USSR) First three-man orbital mission; V. Kamarov, K. Feoktistov, B. Yegerav; 16 orbits; 24 hours 18 minutes.

NOV. 28 ● MARINER 4 (USA) First successful fly-by of Mars.

1965

MAR. 18 ● VOSKHOD II (USSR) First space walk; A. Leonov (performed 10-minute EVA). P. Belyayev; 17 orbits; 26 hours.

MAR. 23 ● GEMINI 3 (USA) First American two-man crew; first manned orbital maneuvers; Gus Grissom, John Young; three orbits; 4 hours 54 minutes.

JUNE 3 ● GEMINI 4 (USA) First American space walk; Edward White (performed 21-minute EVA); James McDivitt; 62 orbits; 4 days 1 hour 54 minutes.

AUG. 21 ● GEMINI 5 (USA) First extended manned flight; Gordon Cooper, Charles Conrad; 128 orbits; 7 days 22 hours.

DEC. 4 ● GEMINI 7 (USA) First rendezvous in space (with Gemini 6); Frank Borman, James Lovell; record 220 orbits; 13 days 18 hours 35 minutes.

DEC. 15 ● GEMINI 6 (USA) First rendezvous in space (with Gemini 7); Walter Schirra, Thomas

Stafford; 16 orbits; 25 hours 51 minutes.

1966

JAN. 31 ● LUNA 9 (USSR) First soft landing on moon; returned lunar-surface photos.

MAR. 16 ● GEMINI 8 (USA) First docking in space with previously launched target (Agena 8); malfunction caused mission to be curtailed; Neil Armstrong, David Scott; 6.5 orbits; 10 hours 41 minutes.

MAR. 31 ● LUNA 10 (USSR) First lunar orbiter; returned lunar data until May 1966.

MAY 30 ● SURVEYOR 1 (USA) First American soft landing on moon; returned 11,240 photos.

JUNE 3 ● GEMINI 9 (USA) Rendezvoused with Agena target; 2-hour-8-minute EVA carried out; Thomas Stafford, Eugene Cernan; 47 orbits; 3 days.

JULY 18 ● GEMINI 10 (USA) First use of target vehicle as source of propulsion after rendezvous and docking; first double rendezvous (with Agena 10 and Agena 8); first retrieval of space object (test package on target vehicle) during EVA; John Young, Michael Collins; 43 orbits; 2 days 22 hours 48 minutes.

AUG. 10 ● LUNAR ORBITER 1 (USA) Orbited moon and returned 207 photos of lunar equatorial region as part of program to aid selection of landing sites for later missions; all orbiters were deliberately crashed on the moon so that their radio transmitters would not interfere with later spacecraft.

SEPT. 12 ● GEMINI 11 (USA) Rendezvous and docking achieved on first revolution; used Agena 11 propulsion to achieve record altitude of 850 miles; Charles Conrad, Richard Gordon; 44 orbits; 2 days 23 hours 18 minutes.

NOV. 11 ● GEMINI 12 (USA) Final Gemini mission; three EVAs, for record total of 5 hours 30 minutes; James Lovell, Buzz Aldrin; 63 orbits; 3 days 22 hours 36 minutes.

1967

JAN. 27 ● APOLLO 1 (USA) Fire inside spacecraft during ground testing resulted in death of three astronauts; Gus Grissom, Edward White, Roger Chaffee.

APR. 23 ● SOYUZ 1 (USSR) First manned test flight of new Soyuz spacecraft; V. Komarov killed on ground impact when parachute lines of reentry module were fouled.

JUNE 12 ● VENERA 4 (USSR) First successful probe of atmosphere on planet Venus.

AUG. 1 ● LUNAR ORBITER 5 (USA) Photographed five potential Apollo landing sites; ended lunar orbiter mapping program.

1968

SEPT. 15 ● ZOND 5 (USSR) First circumlunar flight to return to earth; carried live organisms; recovered in Indian Ocean.

OCT. 11 ● APOLLO 7 (USA) First American three-man earth-orbital mission; Wally Schirra, Donn Eisele, Walt Cunningham; 163 orbits; 10 days 20 hours.

OCT. 25 ● SOYUZ 2 (USSR) Unmanned satellite; rendezvous target for Soyuz 3.

OCT. 26 ● SOYUZ 3 (USSR) Manned spacecraft; maneuvered to within 650 feet of Soyuz 2; G. Beregovoi; 64 orbits; 3 days 22 hours 54 minutes.

DEC. 21 ● APOLLO 8 (USA) First manned orbit of moon; Frank Borman, James Lovell, William Anders; 10 lunar orbits; 6 days 3 hours in space overall.

1969

JAN. 14 ● SOYUZ 4 (USSR) First docking (with Soyuz 5) of two manned Soviet spacecraft; V. Shatalov; 45 orbits; 2 days 23 hours.

JAN. 15 ● SOYUZ 5 (USSR) Docked with Soyuz 4; B. Volyanov, A. Yeliseyev, Y. Khronov; 46 orbits; 3 days.

MAR. 3 ● APOLLO 9 (USA) First test of lunar module in earth orbit; James McDivitt, David Scott, Russell Schweickart; 151 orbits; 10 days 1 hour.

MAY 18 ● APOLLO 10 (USA) First test of lunar module in lunar orbit; Thomas Stafford, Eugene Cernan, John Young; 31 lunar orbits; 8 days in space overall.

JULY 16 ● APOLLO 11 (USA) First manned lunar landing; first men to walk on moon's surface; Neil Armstrong, Buzz Aldrin, Michael Collins; 22 hours on moon, with 2-hour-35-minute EVA.

OCT. 11 ● SOYUZ 6 (USSR) First triple launch (with Soyuz 7 and 8) of manned craft; nondocking group flight; G. Shonin, N. Kubasov; 79 orbits; 4 days 22 hours 42 minutes.

OCT. 12 ● SOYUZ 7 (USSR) Target vehicle for group rendezvous (with Soyuz 6 and 8); A. Filipchenko, V. Volkov, V. Gorbatko; 79 orbits; 4 days 22 hours 42 minutes.

OCT. 13 ● SOYUZ 8 (USSR) Flagship in maneuvers (with Soyuz 6 and 7); V. Shatalov, A. Yeliseyev; 79 orbits; 4 days 22 hours 42 minutes.

NOV. 14 ● APOLLO 12 (USA) Second manned lunar landing; returned parts from Surveyor 3; Pete Conrad, Richard Gordon, Alan Bean; 32 hours on moon, with 7 hours 45 minutes of EVA.

1970

APR. 11 ● APOLLO 13 (USA) Third manned lunar landing attempt, aborted due to oxygen-tank explosion in service module; crew returned safely; James Lovell, Fred Haise, Jack Swigert; 5 days 22 hours 53 minutes.

JUNE 2 ● SOYUZ 9 (USSR) Set new duration record for manned space flight; V. Sevastianov, A. Nikolayev; 268 orbits; 17 days 16 hours 59 minutes.

AUG. 17 ● VENERA 7 (USSR) Venus atmosphere probe; first successful landing on surface.

1971

JAN. 31 ● APOLLO 14 (USA) Third manned lunar landing; collected 96 pounds of lunar samples; Alan Shepard, Stuart Roosa, Edgar Mitchell; 34 hours on moon, with 9 hours 24 minutes of EVA.

APR. 19 ● SALYUT 1 (USSR) Unmanned prototype for orbiting space station and laboratory;

decayed October 11, 1971.

APR. 23 ● *SOYUZ 10 (USSR)* First crew to dock with orbiting *Salyut 1* (5 hours 30 minutes); V. Shatalov, A. Yeliseyev, N. Rukavishnikov.

MAY 30 ● *MARINER 9 (USA)* First successful Mars orbiter; returned 7,000 pictures of surface and moons.

JUNE 6 ● *SOYUZ 11 (USSR)* First crew to occupy orbiting Salyut space station (22 days); because of accidental depressurization, cosmonauts died on reentry; G. Dobrovolsky, V. Volkov, V. Patsayev.

JULY 26 ● *APOLLO 15 (USA)* Fourth manned lunar landing; first use of manned Lunar Roving Vehicle; David Scott, Alfred Worden, James Irwin; 67 hours on moon, with 18 hours 35 minutes of EVA.

1972
MAR. 2 ● *PIONEER 10 (USA)* First successful fly-by of Jupiter; first probe to escape solar system.

APR. 16 ● *APOLLO 16 (USA)* Fifth manned lunar landing; collected 213 pounds of lunar samples; Charles Duke, Ken Mattingly, John Young; 71 hours on moon, with 20 hours 15 minutes of EVA.

DEC. 7 ● *APOLLO 17 (USA)* Sixth and last Apollo manned lunar landing; collected 243 pounds of lunar samples; Eugene Cernan, Ronald Evans, Harrison Schmitt; record 76 hours on moon, with a total 23 hours 12 minutes of EVA.

1973
APR. 5 ● *PIONEER 11 (USA)* Jupiter probe; first successful fly-by of Saturn.

MAY 14 ● *SKYLAB (USA)* First unmanned American space station; placed in earth orbit but damaged during launch.

MAY 25 ● *SKYLAB 1 (USA)* First crew to occupy Skylab (28 days); replaced thermal shield and repaired solar wing; Pete Conrad, Joseph Kerwin, Paul Weitz.

JULY 28 ● *SKYLAB 2 (USA)* Second crew to occupy Skylab (59.5 days); Alan Bean, Jack Lousma, Owen Barriott.

SEPT. 27 ● *SOYUZ 12 (USSR)* Test of modified spacecraft (chemical batteries replace extensible solar panels) for ferry missions to orbiting Salyut laboratories; V. Lazarev, O. Makarov.

NOV. 3 ● *MARINER 10 (USA)* First craft to use gravity of one planet (Venus) to reach another (Mercury); first TV pictures of Mercury.

NOV. 16 ● *SKYLAB 3 (USA)* Third and last crew to occupy Skylab (84 days); longest Skylab mission; Gerald Carr, Edward Gibson, William Pogue.

DEC. 18 ● *SOYUZ 13 (USSR)* Manned spacecraft; astrophysical, biological and earth-resources experiments; P. Klimuk, V. Lebedev; 7 days 20 hours 55 minutes.

1974
JUNE 24 ● *SALYUT 3 (USSR)* Unmanned space station; in orbit 214 days.

JULY 3 ● *SOYUZ 14 (USSR)* Manned spacecraft; crew rendezvoused and docked with *Salyut 3* (14.5 days); P. Popovich, Y. Artyukhin.

AUG. 26 ● *SOYUZ 15 (USSR)* Manned spacecraft; failed to dock with *Salyut 3*; G. Sarafanov, L. Demir.

DEC. 2 ● *SOYUZ 16 (USSR)* Manned spacecraft; tested new design for joint US-USSR Apollo-Soyuz Test Project; A. Filipchenko, N. Rukavishnikov.

DEC. 26 ● *SALYUT 4 (USSR)* Unmanned space station; in orbit 770 days.

1975
JAN. 11 ● *SOYUZ 17 (USSR)* Manned spacecraft; crew docked with and occupied *Salyut 4* (28 days); A. Gubarev, G. Grechko.

MAY 24 ● *SOYUZ 18 (USSR)* Second crew to occupy *Salyut 4* (63 days); P. Klimuk, V. Sevastyanov.

JUNE 8 ● *VENERA 9 (USSR)* Venus probe; soft-landed and returned first TV pictures from surface.

JULY 15 ● *APOLLO-SOYUZ (USA-USSR)* First cooperative international flight; docked in earth orbit for 2 days; T. Stafford, V. Brand, D. Slayton (USA); A. Leonov, V. Kubasov (USSR).

AUG. 20 ● *VIKING 1 (USA)* Mars orbiter; first successful Mars landing; returned photographs and data.

SEPT. 9 ● *VIKING 2 (USA)* Mars orbiter; landed; returned pictures and data.

NOV. 17 ● *SOYUZ 20 (USSR)* Unmanned spacecraft; ground-controlled docking with *Salyut 4* in rehearsal for station resupply.

1976
JUNE 22 ● *SALYUT 5 (USSR)* Unmanned space station; in orbit 412 days.

JULY 6 ● *SOYUZ 21 (USSR)* Manned spacecraft; docked with *Salyut 5*; crew performed extensive experiments; B. Volynov, V. Zholobov; 49 days 6 hours 24 minutes.

SEPT. 15 ● *SOYUZ 22 (USSR)* Manned spacecraft; carried multispectral camera; V. Bykovsky, V. Aksyonov; 7 days 21 hours 54 minutes.

OCT. 14 ● *SOYUZ 23 (USSR)* Manned spacecraft; failed to dock with *Salyut 5*; V. Zudov, V. Rozhdestvensky.

1977
FEB. 7 ● *SOYUZ 24 (USSR)* Manned spacecraft; docked with *Salyut 5*; tested and repaired parts aboard space station; V. Gorbatko, Y. Glazkov; 17 days 16 hours 8 minutes.

AUG. 20 ● *VOYAGER 2 (USA)* Fly-by probe of Jupiter, Saturn, Uranus, Neptune; returned pictures and data.

SEPT. 5 ● *VOYAGER 1 (USA)* Fly-by probe of Jupiter and Saturn; returned pictures and data on Jupiter and five of its moons and on Saturn and four of its moons.

SEPT. 29 ● *SALYUT 6 (USSR)* Unmanned space station; sent into earth orbit.

OCT. 9 ● *SOYUZ 25 (USSR)* Manned spacecraft; failed to dock with *Salyut 6*; V. Kovalenok, V. Ryumin.

DEC. 10 ● *SOYUZ 26 (USSR)* Manned spacecraft; docked with *Salyut 6*; 1-hour-28-minute EVA; Y. Romanenko, G. Grechko; 96 days 10 hours.

1978
JAN. 10 ● *SOYUZ 27 (USSR)* Manned spacecraft; docked with *Salyut 6*, achieving first three-spacecraft complex (with *Soyuz 26*); V. Dzhanibekov, O. Makarov; 6 days.

JAN. 20 ● *PROGRESS 1 (USSR)* Unmanned, expendable transport craft to resupply *Salyut 6* with propellants, food and other cargo; on February 7 it was made to reenter the atmosphere and burn up.

MAR. 2 ● *SOYUZ 28 (USSR)* Manned spacecraft; docked with *Salyut 6*; first international crew to occupy *Salyut 6*; A. Gubarev, V. Remek (Czechoslovakia); 7 days 20 hours.

JUNE 15 ● *SOYUZ 29 (USSR)* Manned spacecraft; docked with *Salyut 6*; performed 2-hour EVA to replace equipment and retrieve package exposed to space for 10 months; V. Kovalenok, A. Ivanchenkov; 139 days.

JUNE 27 ● *SOYUZ 30 (USSR)* Manned spacecraft; international crew; docked with *Salyut 6*; performed biomedical experiments; P. Klimuk, M. Hermaszewski (Poland); 7 days 22 hours.

AUG. 8 ● *PIONEER-VENUS 2 (USA)* Five entry probes measure Venus' atmosphere before landing; returned surface data.

AUG. 26 ● *SOYUZ 31 (USSR)* Manned spacecraft; international crew; docked with *Salyut 6*; V. Bykovsky, S. Jaehn (East Germany); 7 days 20 hours 49 minutes.

1979
FEB. 25 ● *SOYUZ 32 (USSR)* Seventh crew to occupy *Salyut 6*; deployed radio telescope; V. Lyakhov, V. Ryumin; 175 days.

APR. 10 ● *SOYUZ 33 (USSR)* Manned spacecraft; failed to dock with *Salyut 6*; N. Rukavishnikov, G. Ivanov.

JUNE 6 ● *SOYUZ 34 (USSR)* Unmanned spacecraft; ground-controlled docking with *Salyut 6* and *Soyuz 32*; returned on August 19 with crew from *Soyuz 32*.

DEC. 16 ● *SOYUZ T-1 (USSR)* Unmanned, new-generation Soyuz craft; ground-controlled docking with *Salyut 6*.

1980
APR. 9 ● *SOYUZ 35 (USSR)* Manned spacecraft; eighth crew to occupy *Salyut 6*; L. Popov, V. Ryumin; 184 days 20 hours 12 minutes.

MAY 26 ● *SOYUZ 36 (USSR)* Manned spacecraft; international crew; docked with *Salyut 6*; V. Kubasov, B. Farkas (Hungary); 7 days 20 hours 46 minutes.

JUNE 5 ● *SOYUZ T-2 (USSR)* First manned flight of the Soyuz T series; docked with *Salyut 6*; Y. Malyshev, V. Aksyonov; 3 days 22 hours.

JULY 23 ● *SOYUZ 37 (USSR)* Manned spacecraft; international crew; docked with *Salyut 6*;

performed a series of experiments with resident crew (Popov and Ryumin); V. Gorbatko, P. Tuan (Vietnam); 7 days 20 hours 42 minutes.

SEPT. 18 ● *SOYUZ 38 (USSR)* Manned spacecraft; international crew; ferried fuel and supplies to *Salyut 6*, and returned *Soyuz 35* crew; Y. Romanenko, A. Mendez (Cuba); 7 days 20 hours 43 minutes.

NOV. 27 ● *SOYUZ T-3 (USSR)* Manned spacecraft; first Soyuz in nine years to carry three crew members; L. Kizim, G. Strekalov, O. Makarov; 15 days.

1981
MAR. 12 ● *SOYUZ T-4 (USSR)* Manned spacecraft; docked with *Salyut 6*; carried out repairs to space station; V. Kovalenok, V. Savinykh; 76 days.

MAR. 22 ● *SOYUZ 39 (USSR)* Manned spacecraft; international crew; docked with *Salyut 6*; first TV transmission from space of holographic images; V. Dzhanibekov, J. Gurragcha (Mongolia); 9 days.

APR. 12 ● *STS-1 (USA)* First orbital test flight of space shuttle *Columbia*; tested cargo-bay doors; John Young, Robert Crippen; 2 days 6 hours.

MAY 14 ● *SOYUZ 40 (USSR)* Manned spacecraft; docked with *Salyut 6*; studied effects of space on construction materials; L. Popov, D. Prunariu (Rumania); 9 days.

NOV. 12 ● *STS-2 (USA)* Second orbital test flight of space shuttle *Columbia*; first in-flight test of manipulator arm; Joseph Engle, Richard Truly; 2 days 6 hours.

1982
MAR. 22 ● *STS-3 (USA)* Third orbital test flight of space shuttle *Columbia*; first manipulation of payload in cargo bay; Jack Lousma, Gordon Fullerton; 8 days.

APR. 19 ● *SALYUT 7 (USSR)* Unmanned space station; test of systems and equipment for future crew occupation.

JUNE 24 ● *SOYUZ T-6 (USSR)* Manned spacecraft; international crew; docked with *Salyut 7*; V. Dzhanibekov, A. Ivanchenkov, J. Chrétien (France); 7 days

JUNE 27 ● *STS-4 (USA)* Fourth orbital test of space shuttle *Columbia*; first commercial experiments and "getaway specials"; Ken Mattingly, Henry Hartsfield; 8 days.

AUG. 19 ● *SOYUZ T-7 (USSR)* Manned spacecraft; first coed crew (2 male, 1 female); docked with *Salyut 7*; L. Popov, A. Serebov, S. Savitskaya; 9 days.

NOV. 11 ● *STS-5 (USA)* Fifth flight of space shuttle *Columbia*; first operational manned shuttle flight; launched a pair of commercial satellites; Vance Brand, Robert Overmyer, William Lenoir, Joseph Allen; 5 days.

1983
APR. 4 ● *STS-6 (USA)* First flight of space shuttle *Challenger*, second shuttle in US fleet; deployed tracking and data-relay satellite; Paul Weitz, Karol Bobko, Story Musgrave, Donald Peterson; 5 days.

LIFE IN SPACE

LITTLE, BROWN AND COMPANY   BOSTON   TORONTO

B

LIFE IN SPACE
EDITOR: Robert Grant Mason
Deputy Editor: Roberta R. Conlan
Designers: Ray Ripper, Edward Frank
(principals), Ellen Robling
Chief Researchers: Patti H. Cass,
W. Mark Hamilton
Staff Writers: Donald Davison Cantlay,
Thomas H. Flaherty Jr., John Manners
Researchers: Loretta Britten,
M. Barbara Brownell, Janet Doughty,
Adrianne T. Goodman, Marta A. Sanchez
Copy Coordinator: Barbara Fairchild Quarmby
Art Assistants: Kenneth E. Hancock,
Lorraine D. Rivard, Peter Simmons
Picture Coordinator: Renée DeSandies
Editorial Assistant: Carolyn Wall Halbach
Special Contributors: Oliver Allen,
Ronald H. Bailey, Rachel S. Cox, Don Earnest,
David Friend, Leon Greene, Charles Osborne,
Charles Smith, Keith Wheeler

Editorial Operations
Design: Anne B. Landry (art coordinator),
James J. Cox (quality control)
Research: Jane Edwin (assistant director),
Louise D. Forstall
Copy Room: Diane Ullius (director),
Celia Beattie
Production: Feliciano Madrid (director),
Gordon E. Buck, Peter Inchauteguiz

Correspondents: Elisabeth Kraemer (Bonn); Margot
Hapgood, Dorothy Bacon (London); Miriam
Hsia, Lucy T. Voulgaris (New York); Maria Vincenza
Aloisi, Josephine du Brusle (Paris); Ann
Natanson (Rome). Valuable assistance was also
provided by: John Dunn (Melbourne); Christina
Lieberman, Cornelis Verwaal (New York).

CONSULTANTS
GREGORY P. KENNEDY, co-author of the *Space
Shuttle Operator's Manual*, a layman's guide to
piloting the shuttle, is assistant curator of the Space
Science and Exploration Department of the
National Air and Space Museum in Washington,
D.C. He coordinates the museum's space and
rocket exhibits and is responsible for its collection
of space artifacts.

DR. TED A. MAXWELL is chairman of the
National Air and Space Museum's Center for Earth
and Planetary Studies. A widely published
specialist in planetary geology, he also serves as
director of the museum's Regional Planetary
Image Facility and as curator of its Exploring the
Planets Gallery.

Library of Congress Cataloguing in Publication Data
Main entry under title:
Life in space.
  Bibliography: p.
  Includes index.
  1. Astronautics — History — Pictorial works.
I. Life (Chicago)   II. Time-Life Books.

Published by arrangement with Time-Life Books

Published simultaneously in Canada
by Little, Brown & Company (Canada) Limited

PRINTED IN THE UNITED STATES OF AMERICA

ON THE COVER:
Reflecting the lunar module and faraway
earth in his visor, astronaut Russell Schweickart
carries a 70mm camera while on a space walk
to test his life-support backpack on March 6,
1969. It was the fourth day of *Apollo 9's*
10-day mission—the next-to-last rehearsal for
Apollo 11's landing on the moon.

The events shown and described in this book have taken place in little more than 25 years, only a wink of time in the sweep of history. Yet their significance is enormous, for in this period we have first reached beyond the planet that has been home since the beginning. These years have seen the start of our greatest adventure, the exploration of space, which has already progressed from cautious ballistic hops with animal riders, to manned landings on the surface of the moon, to probes of deep space that have even passed the orbit of Neptune, 2.8 billion miles from a comforting sun, on flights to heaven knows where.

What our accomplishments will be in the next 25 years, or 100, no one can really say. But this much is certain: The record of the opening era of exploration is extraordinary in its completeness and diversity. That the record is also spectacular and often moving can be credited to the imagination and skill of the participants in the programs, the technicians and the astronauts themselves—and to the ingenuity of the photographers and reporters who covered these events as they took place. Without question, foremost among these were the journalists from *Life*.

Knowing what was happening behind the scenes on a day-to-day basis in the Mercury, Gemini and Apollo programs was the preoccupation of a number of *Life* staffers over many years. The magazine did have a contract with the astronauts and their families for their exclusive personal accounts of the flights, but to cover the whole manned space story required a task force.

Life's Military Affairs Editor, John Dille, directed much of the coverage under Managing Editor Edward K. Thompson. Later Gemini and Apollo stories were produced under Managing Editors George Hunt and Ralph Graves, who was the boss at *Life* at the time of the first moon landing. Some of the best early reporting and writing were contributed by Don Schanche, Richard Billings and Ron Bailey who, like other members of the astronaut-coverage group, devoted all their working time to the project.

One of the several *Life* journalists who took on the assignment with barely more than a passing acquaintance with space was Dora Jane Hamblin. Dodie Hamblin went to Houston to work on the Apollo program in the late 1960s; and by the time of the moon landing in 1969, she was expert enough to become the co-author (with *Life's* Gene Farmer) of the most comprehensive of the *Apollo 11* books, *First on the Moon*. Hamblin came to know the wives of the moon landing crew particularly well in that period when the whole country was looking at them on television screens. "Pat Collins rolls with the punches," she once advised her editors in a quick briefing. "Joan Aldrin is emotional. Jan Armstrong is uptight." Then Hamblin added by way of further guidance, "You have to watch their hands, their toes, to catch their moods. If you watch their faces, they'll just smile."

But the *Life* staffer who, more than any journalist, devoted himself completely to the business of documenting America's thrust into space was photographer Ralph Morse. Long before Morse became fascinated with space, he had a reputation for being an extraordinarily versatile photographer, capable of first-rate performance on assignments that demanded ingenuity and technical creativity or on stories that required a more candid and intimate approach.

A New Yorker by birth, the short, fast-talking, marvelously alert and energetic Morse was taking pictures professionally by the time he was 20, and he joined the staff of *Life* in 1942, when he was just 24. Immediately he became a combat photographer, one of the best. He worked extensively throughout the War in both the Pacific and European theaters and was commended for bravery under fire by U.S. military officers in France.

Morse was one of a handful of correspondents who covered the Marine landing in Guadalcanal in 1942, but his pictures of that invasion were never published. The pictures, his cameras and $400 in cash he later claimed quite justly as a legitimate expense item were lost when the ship Morse was on, the heavy U.S. cruiser *Vincennes,* was sunk along with three others in a lightning night attack by the Japanese off Savo Island in the Solomons. The unsinkable Morse spent the six and a half dark hours before rescue keeping himself and a young Naval officer afloat in a single life ring.

For his coverage of the various astronaut programs, Morse made it his business to get to know well all the pilots, their wives and their children. He fished, sailed and water-skied with his subjects and developed a genuine closeness to them. He studied all the training devised so that he could decide how to shoot pictures that no one else would think of. When special equipment was needed to get the shot, he designed and built it.

Morse understood enough about the plans for the great variety of missions that on numerous occasions he was able to suggest to the flight directors and the astronauts how to take photographs to improve their own record of the flights. He all but became a member of this select group of pilots, and their jokes about his persistence and his enthusiasm were expressions of their real admiration for him.

Often Morse found out more about the pilots than they wanted him to know. The press was barred from one Mercury training exercise in 1960; the astronauts were going to practice survival techniques to be used in the unlikely event that the capsule could not make a water landing but was forced down in some desert area instead. Naturally, the prospect of seeing his friends wading around in the hot sand was irresistible to Morse. He knew that the exercise would take place in the vicinity of Reno, Nevada, so he flew there a few days before the pilots were scheduled to arrive.

Next, Morse chartered a small plane and scouted the surrounding area for miles in every direction, looking for signs. He spotted what looked like a freshly set-up marker for a possible helicopter landing area. Morse had his pilot fly from there to the nearest highway, where he dropped three flour sacks on the side of the road. Then they flew back to Reno, where Morse rented a Jeep, drove out the highway to the site of his flour bombing, and hiked in to the spot where he had discovered the helicopter marker. He was ready and waiting for the astronauts when they arrived for the exercise *(page 30)*—just as John Glenn had predicted he would be.

To reflect on *Life's* coverage of those years—especially the years of the first astronaut flights—is to recall the way we felt about the events we were witnessing. We were thrilled by them—caught up in the currents of excitement and renewed pride experienced by most Americans, who had been suffering from the pervasive national sense of ineptitude and gloom that followed the Soviet triumph with *Sputnik 1* in 1957. It felt great to see those early boosters fly. I can remember standing on the beach at Cape Canaveral in 1961 with Scott Carpenter's family and watching with tears in my eyes until the Mercury capsule with Ham, the chimp, aboard passed out of sight downrange—and we all reveled in the sense of being on a winning team.

Our attitude may not have made for the most skeptical and objective journalism; to some extent we were too passionately involved for that. And why not? The astronauts were more or less the same age as we were, and they'd been preselected to become heroes in a great national effort. It was pretty heady stuff, and it would have taken the sternest sort of self-discipline or a great lack of imagination for a reporter meeting regularly with the pilots not to fantasize a piece of the adventure for himself. At least, that was the way I found myself occasionally reacting, all at the same time that I was quite sure I was not made of the right stuff.

Several of the pilots were generous about sharing with us what they could of their amazing work and their own feelings about it. Deke Slayton, Carpenter and Glenn, without saying anything about it, intuitively understood the vicarious nature of our enthusiasm. After Glenn flew his historic orbital mission in February of 1962, after he had been decorated by President John F. Kennedy and had made an address to Congress in Washington, he went to New York with all the Mercury astronauts and their families for what was the biggest ticker-tape parade ever. The crowds along the route were so great that the fenders of the Cadillac carrying Glenn and his family were dented in the crush.

When the parade was over, the hero from New Concord, Ohio, and his immediate entourage went back to his suite at the Waldorf Astoria Towers, where they planned to have a quiet family dinner before going to the theater. Glenn's parents and the parents of his wife, Annie, were there, too, along with a few close friends and some security men. I had been invited as well; I was very eager to go over with Glenn the first-person account of the flight I had gotten from him the day before in Washington.

Just before dinner, the astronaut suddenly decided that he needed a haircut. A barber was summoned to the suite, and a straight chair in the living room was draped for the occasion. While Glenn's already short red hair was being cut, he and I, sitting on a hassock next to him, worked on the manuscript of the interview.

When the barber was finished, Glenn stared at me for a moment or two before getting out of the chair. "You sure could use a trim, too, Loudon," he said, getting up. "Be my guest." We exchanged seats, and with me then under the scissors, we went on with the editing.

In the early days of the Mercury program, much was made of the extraordinary talents of the astronauts—their brightness, their adaptability, their physical conditioning. By advance reputation they seemed almost superhuman. Yet as far as their feelings went, they turned out to be refreshingly like the rest of us. Just a few days after Alan Shepard made the historic first U.S. manned flight into space, in May 1961, I met him alone at his house near Norfolk, Virginia, to get his first-person account of the mission.

Of all the seven Mercury astronauts, Shepard had the reputation of being the most cool and self-possessed. Nothing much fazed him. Under stress his emotions were perfectly contained, we were told by his fellow pilots. Now, in the comfort of his own backyard a couple days after the mission, he maintained coolly that the ride had been pretty much as he expected. He'd been well trained, he said, to deal with any emergency on the 15-minute, 115-

mile-high trip; he didn't take up any of his valuable time with worrying. A piece of cake, he assured me.

Yet for all his protestations of being absolutely at ease with the whole thing, Shepard's voice on the tape told me something quite different. As I listened to it again and again over the next few hours preparing the story, I could hear the unmistakable sound, amid the domestic noises of dogs barking and neighboring lawns being cut, of Shepard's voice trembling as he recounted the details of the lift-off. There was simply no doubt that this laconic Navy test pilot from New Hampshire, the first American to have heard a rocket rumble into life beneath him seconds before it hurled him into space, was moved again just telling about it.

More than 20 years have passed since that moment, and the parade of big events in space has gone on unabated. The detailed and spectacular coverage, much of it contributed by photographers from NASA, has continued, too. Even during the years between the end of the weekly *Life* in 1972 and the start of the monthly magazine in 1978, Ralph Morse went on producing his historic pictures for *Time.*

The best of this superb record, along with *Life's* documentation of the years just before the manned flights began and the later coverage of the space shuttle program and the probes into deep space, finds a new currency in the pages of this book, produced with 55,000 words of new text by the editors of Time-Life Books. Taken altogether, it is a fresh and exciting celebration of the peak moments of a truly astounding human journey. □

Floating weightless in the cabin of an Air Force C-135, *Life* **photographer Ralph Morse and a reporter undergo one of the esoteric sensations experienced by astronauts in training.**

MERCURY'S BOLD SEVEN

They sat enthroned on a dais in Washington, D.C., seven smiling, seemingly relaxed young Americans, confronting a roomful of clamoring reporters and photographers. It was April 9, 1959. A year and a half earlier, the nation's self-esteem had been dealt a stunning blow: With the launch of *Sputnik I (page 12)*— and of *Sputniks II* and *III* in frighteningly short order—the Soviet Union had cleared the first hurdles in the space race, leaving the United States still at the starting line. For months the country had reeled under the shock, worrying about Soviet missiles, fretting over school curriculums in anxiety to produce better scientists, and chafing at having been proved second best.

Now here were seven men committed to redressing the nation's humiliation. Donald (Deke) Slayton, Alan Shepard, Walter (Wally) Schirra, Virgil (Gus) Grissom, John Glenn, Leroy (Gordon) Cooper, Malcolm (Scott) Carpenter: These were the astronauts of Project Mercury. One of them, it was fervently hoped, would become the first human to leave the earth and return to tell about it. The others would follow, each charting space in his turn.

Outwardly the seven seemed to be much alike. They were all military test pilots: three Air Force, three Navy and one Marine. President Dwight D. Eisenhower had vetoed an early scheme to experiment with scuba divers, mountain climbers—almost anybody with a proven penchant for high adventure.

All were under 40; youth would count. Each weighed in at 180 pounds or less; extra pounds would be a curse aboard the vehicle intended to rocket them into space. None was taller than 5 feet 11 inches; the cramped tin can that would contain them had no more room for more than 71 inches of humanity at full stretch. All were married; all had children. All had lived the peripatetic life of professional service personnel. Each

had proved himself hard to scare.

Now they coalesced into a public image of the seven musketeers, one for all and all for one. In truth, however, the astronauts were alike in one more significant detail: Each was fiercely competitive and ambitious, primed to win the professional immortality of being first in space—and never mind the price that might be paid. "A seven-sided coin," said Schirra, hinting at their individual yearnings to be first. "The nearest to heaven I will ever get," said Glenn.

In still another aspect they were authentic brothers. Each had been squeezed through the same excruciating wringer. The process of finding them had begun with the service records of 508 military test pilots, which the Space Task Group (STG) at Langley Air Force Base in Virginia weeded down to 110 likely candidates. Out of the first 69 who were summoned for interviews, 32 volunteered—so

Seven astronauts signal an enthusiastic affirmative when asked by a brash journalist whether they really expect to return alive from space. Schirra and Glenn, evidently twice as confident, shot up both left and right hands. From the left: Slayton, Shepard, Schirra, Grissom, Glenn, Cooper and Carpenter.

many that the rest were never called.

From Langley they were shipped off in groups to a civilian clinic in Albuquerque, New Mexico, for a week of extremely rigorous—and often embarrassing—tests. In some 30 different laboratory procedures their bodies were mapped inside and out. Eyes alone underwent 17 different ophthalmologic examinations. The men were dunked to the chin in water to measure their specific gravity. They trudged through the halls in hospital gowns, lugging briefcases in one hand and gallon jugs in the other because the doctors demanded full 24-hour urine samples.

One candidate failed the medical examination. The remaining 31 were shuttled off to the Aeromedical Laboratory at Wright-Patterson AFB in Ohio. To prove that they were the "ordinary supermen" NASA was looking for, they were subjected to increasing degrees of heat, pressure, ice water, noise, vibration, acceleration and disorienting flip-flops on power-actuated tilt tables. They sprinted on treadmills and blew up balloons until their outraged lungs said No more. And in addition to these physical torments, each was also required to spend a week baring his soul to two white-smocked psychologists.

Ultimately, 18 came through the ordeal as certified supermen. STG selected seven of them on the rather unscientific grounds that they seemed likely to get along with one another. Then they were shipped to Washington to see if they could also endure a press conference.

At the time of the Sputniks, the U.S. space program was not so much in disarray as it was in too many contradictory kinds of array—most of them military. The Army was toying with the notion of firing a man atop one of the Redstone missiles developed by Wernher von Braun *(pages 17-19)* and his team of German rocketeers.

The Navy was tinkering with the idea of a cylindrical vehicle that, once in space, would inflate fabric wings and glide down. The Air Force was trying to develop huge rockets and contemplating an orbital bomber to be called Dyna-Soar. The National Advisory Committee for Aeronautics (NACA) was building the X-15 rocket plane and hoping someday to orbit it.

Most of these programs were on the verge of withering away for lack of funds. President Eisenhower believed space exploration should be a peaceful pursuit, and he was against spending vast sums of money on it. Moreover, he was reluctant to send a man up there to do a machine's work.

The Sputniks and alarmed public pressure precipitated an abrupt change. By July 1958, Eisenhower had signed an act transforming NACA into NASA (the National Aeronautics and Space Administration), which was empowered and funded to con-

centrate the effort and to draw on the facilities of the other programs.

In some ways, the new agency could count itself fairly well off. NACA scientists and engineers—most of whom made the transition to NASA—had been looking far into the future and working on techniques to send a man to the moon and back. In addition, a respectable amount of rocket power was at hand in the Army's Redstone, an advanced version of Germany's wartime V-2, with 75,000 pounds of thrust. When more power would be needed, the Air Force Atlas, with a projected 360,000 pounds of thrust, would probably be ready.

But the bigger question was the spacecraft itself. For one thing, an astronaut coming back from space would require protection from the 3,000° F. of heat that the vehicle would encounter when it hit the atmosphere. Gradually an idea and a shape evolved, largely the work of a NASA engineer named Maxime Faget.

Faget's brain child was a squat, funnel-shaped capsule that would sit on the tip of the rocket at takeoff, blunt end down. Released from the booster and into orbit, it would tip over and fly blunt end forward. That end would be blanketed with a shield of resinous material designed to burn off—or ablate—in the heat of reentry; its progressive disintegration would protect the rest of the capsule and its rider from the incinerating heat. Faget and his colleagues are also credited with the decisions to drop the capsule by parachute through its last 21,000 feet of descent and to land it in the sea with an air bag to cushion the impact.

The next great problem was reliability. With more than 40,000 critical parts in the booster and another 40,000 in the capsule, making sure they would all work was a daunting proposition. The nation's record up to mid-1959 left room for scant optimism. While von Braun's team had earned brilliant success with three Explorer satellites in 1958—discov-

ering the Van Allen radiation belt with one, inserting another into orbit for 452 days—the space program had also suffered spectacular failures. An early test Vanguard had blown up on the pad in 1957, *Explorer II* had misfired and crashed into the Atlantic in 1958, and a Jupiter rocket had carried a squirrel monkey to a wet grave near the end of 1958.

The key to reliability appeared to be what the engineers called redundancy. If one computerized black box could tell the capsule when to do a back flip, two black boxes would be better in case the first failed. But there were so many functions that teams of black boxes would be needed, adding to the nearly insoluble puzzles of room and weight. When possible, the engineers miniaturized—but microminiaturization was still in its infancy, and the tiny devices were not altogether trustworthy.

While the engineers grappled with the hardware, the astronauts reported to NASA for what was to be two years of intensive training. Nobody knew exactly what burdens their bodies and minds would face in the unknown environment of space, but the training regimen was designed to anticipate what the engineers and doctors considered likely.

The astronauts rode a giant centrifuge that swung them in dizzy circles until the G forces squeezed their muscles and bones and innards with 16 times normal gravity. They sweated in heat chambers until the doctors knew that these men could withstand 135° F. for as long as two hours. They rode a contraption called the MASTIF (for Multiple Axis Space Test Inertia Facility), which simultaneously spun, swung and somersaulted them at rates up to 30 revolutions per minute. The idea was to discover how soon they could regain control if their vehicle started to gyrate in space.

NASA planners worried about the effects of weightlessness in orbit, and they did their best to prepare the astronauts. The men were sent out to

Edwards AFB in the California desert to ride fighter jets in high-speed parabolic arcs, at the peak of which they would go feather-light for a few seconds. Later they were trained in larger transport jets, in which they could float around a padded cabin at zero G for as long as 35 seconds at a time. The astronauts repeated these off-worldly sensations again and again, so as to become familiar with the impossible.

At length the astronauts were introduced to the capsule. They quickly retired to brotherly conference—and they emerged with a united front of we-don't-like-its. The mitten-tight canister gave them claustrophobia. They could put up with the size but they had to see out. Two little portholes and a periscope were not enough; they demanded a picture window. The engineers protested that the best window they could devise was likely to rupture in the stress of space. The astronauts wanted it anyhow—and they won.

They also wanted to be able to get out in a hurry. As designed, the capsule hatch was secured with 70 bolts, which took 65 minutes to draw tight—from the outside; in an emergency, the astronauts would be trapped until help arrived. They got a new hatch, fastened with explosive bolts that could be blown from the inside with the punch of a button.

Most of all, the astronauts wanted to be able to fly their vehicle. They were test pilots—not mere "spam in a can." Black boxes backing up black boxes were fine—so long as the last box worked. If it did not—well, the only reason they might survive would be that they had long ago learned how to work themselves out of jams that they or their machinery had got them into.

In the end the astronauts won almost every demand. The black boxes were still there, in duplicate and triplicate. But the astronauts got a number of pilot controls: They could leave the boxes to their own computerized devices, take over on a separate manual system, or go to a semiautomatic fly-by-wire system. On fly-by-wire they could plug into the automatic system, using a three-axis control stick to adjust the capsule's pitch, roll and yaw.

The astronauts' influence on the control system was largely the work of Gus Grissom. Early on, since there was so much to learn and far more to do than any individual could manage, the seven men divided the job into subject areas—each specialist to brief the others. Grissom worked closely with the capsule contractors, flying out to St. Louis to check each new development. Meanwhile, Carpenter took over communications, Glenn the layout of the instrument panel, Schirra the pressure suit, and Cooper and Slayton the liaison with the booster builders.

Alan Shepard had the widest-ranging job of all. He studied the tracking system, which would be essential for monitoring the condition of capsule and pilot at every moment in orbit, all around the world. In the end, the system involved 18 ground stations for radar, telemetry and voice, 177,000 miles of wire, two ships at sea and a cost of $41 million.

The astronauts had almost everything to make them happy, but one thing galled. As test pilots with healthy egos, they hated to be reminded that "a monkey is going to make the first flight." It was true. While the astronauts busied themselves preparing for space, veterinarians in New Mexico were grooming 20 chimpanzees (pages 33-37) to do the same thing sooner. The chimps were fast learners. A particularly bright chimp named Ham flew up and splashed down successfully on January 31, 1961, and one called Enos went into orbit 10 months later.

All this toil was aimed at one supreme goal—putting a human being into space and, if possible, doing it first. On February 21, 1961, NASA announced that one of three men, Glenn, Grissom or Shepard, would become the first American in space. But on April 12 the Russians scooped the United States again—Major Yuri Alekseyevich Gagarin vaulted aloft and into one full orbit around earth (pages 38-40).

It was a long time coming, but the American astronauts' day did at last dawn. On the morning of May 5, 1961, Lieutenant Commander Alan Shepard, 37, lay atop a Redstone rocket six stories high at Cape Canaveral, Florida. He lay on his back, his body clasped in a foam-lined couch cast to fit him. He had been up since 1 a.m., inside the capsule since 5:20, and he was growing a little impatient as the countdown ran into one hold after another. He spoke into the microphone connecting him to the launch crew on the ground: "Why don't you fix your little problems and light this candle?"

Finally the umbilical cord connecting the capsule to power from the ground fell away. The periscope retracted, and Shepard felt the butterflies in his stomach start up. "Okay, buster, you volunteered for this thing," he told himself. "Now it's up to you to do it." Then he felt and heard the rumble of the engines starting beneath him and there was no more time for butterflies. It was 9:34, and Freedom 7 was moving.

Lift-off was considerably smoother than he had expected—gradual, even gentle. Then, about a minute into the rocket burn, the ride suddenly turned rough—the Redstone seemed to be trying to shake itself apart. Shepard made the decision not to report it. "I did not want to panic anyone into ordering me to leave," he explained later. He waited until the vibration stopped, and then let ground control know indirectly by reporting that things were "a lot smoother now, a lot smoother."

By three minutes into the flight, Freedom 7 was free of the booster and the capsule had about-faced to the heat-shield-forward position. Shepard now had about 12 minutes of flying time left—time to go to work on the prime task of determining whether an astronaut could, in fact, control the spacecraft's attitude. He switched to manual control, taking over one axis at a time, and then took control of all three axes. The feel was reassuringly similar to that on the procedures trainer.

He switched back to auto pilot and let the black boxes run the show, but at the peak of his trajectory, he switched to fly-by-wire and corrected the capsule's pitch. Then the retro-rockets fired with a jolt, and the G forces began building fast toward a high of 11 Gs. Back on automatic control, the capsule's rate of descent was faster than expected, and Shepard waited anxiously for the drogue and main chutes to deploy. At 21,000 feet, the drogue snapped out on time, and at 10,000 feet came the jolt of the great red-and-white main.

The astronaut's debriefing aboard the carrier Lake Champlain was an exercise in jubilation, which was continued by the country as a whole when he reached land again. Shepard had shown no ill effects, had functioned normally in space, had found his five minutes of weightlessness "just a pleasant ride."

Later debriefings led to changes in the capsule's control system and in the pressure suit, all ready in time for the second suborbital flight to be made by Air Force Captain Gus Grissom. The 35-year-old Korean War veteran was eager to fly, but things kept going wrong with the Redstone or with pieces of the capsule.

On July 21, 1961, the countdown finally reached zero and Grissom went up, not experiencing much buffeting going past Mach 1. Up on top he marveled at a black sky with one bright star, was bothered by nothing except a certain stiffness in the manual controls, and splashed down without incident.

While the pickup helicopters converged, he disconnected his oxygen hose, took off his helmet and rolled up the rubber collar that provided a

SPUTNIK: A RUSSIAN COUP

In hindsight, perhaps one of the greatest blessings for the American space program was the launch of the Soviet Union's *Sputnik I* on October 4, 1957. It seemed much otherwise at the time. That 23-inch aluminum ball looping lopsidedly around earth from perigee to apogee offered both shock and indictment. It was an intolerable affront to Western science and galvanized the United States into action.

Looked at one way, *Sputnik I* seemed harmless enough: 184.3 pounds of polished metal peacefully beep-beeping from space. But it was *up* there—visible proof of a frightening Soviet power and technology. It set off endless speculation—some humiliating, some envious, a lot apprehensive. Scientists calculated that it would have taken 400,000 pounds of booster thrust to throw that weight into orbit and somberly acknowledged that the biggest U.S. rocket—the 360,000-pound Atlas—was not yet operational. With all that thrust and guidance, could the Russians not also send a missile accurately half around the world? "A week of shame and danger," lamented Senator Henry Jackson. Russian secretiveness with practical data did nothing to allay the concerns, nor did Nikita Khrushchev's crowing that "Now the bomber and fighter can go into the museums." A month later, *Sputnik II* went up, six times the weight of its predecessor and carrying a dog wired to transmit its vital signs.

The dog died of asphyxiation after a week, and by the time *Sputnik II* had burned in the atmosphere the following April, the United States had three of its own satellites up. But *Explorer I, Vanguard* and *Explorer III* together weighed only 65 pounds, and Khrushchev jeered about American "oranges" in space. On May 15, as if to rub it in, the Russians sent up *Sputnik III*, carrying 2,129 pounds of instruments and circling over most of the earth's inhabited territory.

After the separation, *Sputnik I's* carrier rocket tagged along in slightly lower orbit. Its progress through the sky above Montreal was recorded in time exposure by *Life's* Robert Kelley.

Reflecting the world's awe, a Memphis astronomer proceeds to his observation post laden with the tools of his calling. Millions of others joined in the nighttime vigil as well.

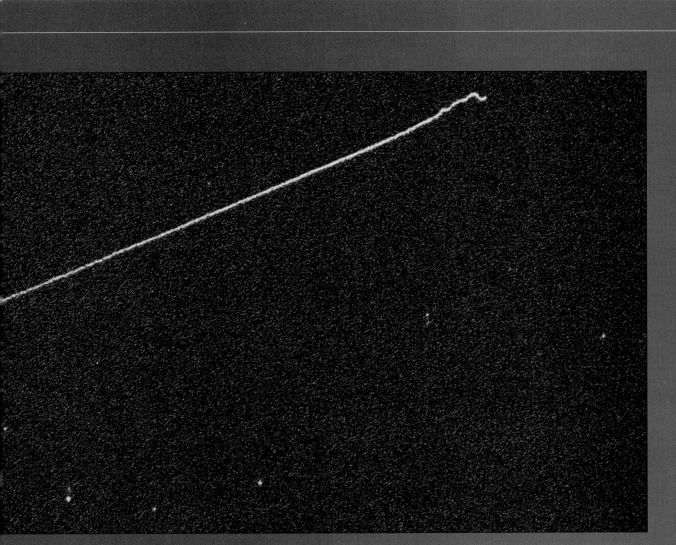

Chagrined or not, no American cartoonist could resist *Sputnik's* temptation. Here in the *Detroit Free Press*, the satellite knocks off Uncle Sam's star-spangled topper.

RUSSIAN SCIENCE

U.S. SATELLITE PROGRAM

Even earthbound, *Sputnik* (meaning "fellow traveler") was a superstar. Here, not long after it was launched, crowds in Moscow gaze with admiration at a replica of the satellite.

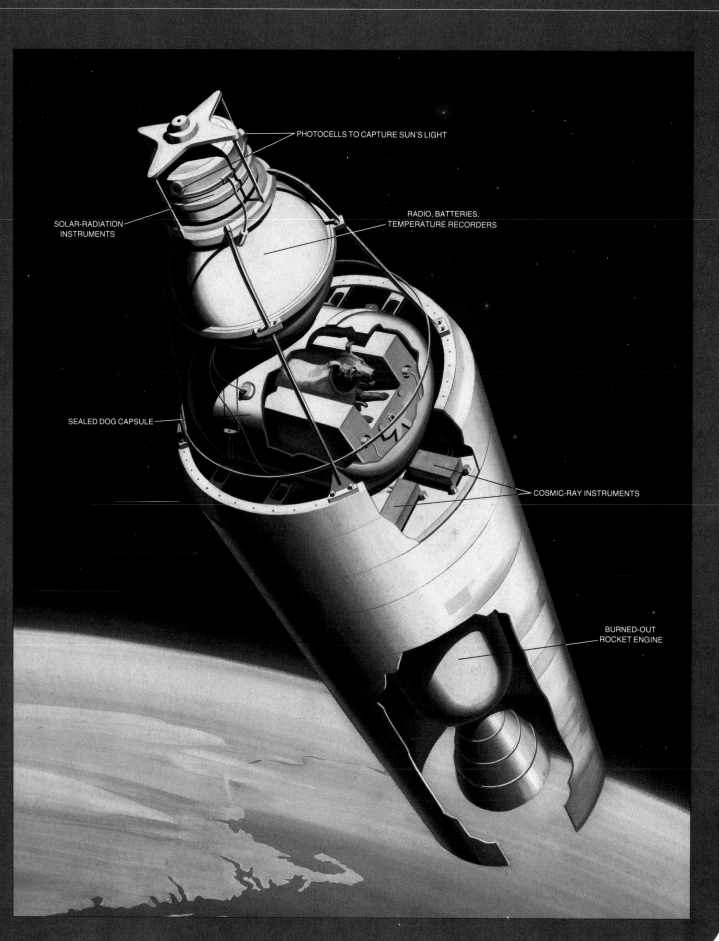

PHOTOCELLS TO CAPTURE SUN'S LIGHT

SOLAR-RADIATION
INSTRUMENTS

RADIO, BATTERIES,
TEMPERATURE RECORDERS

SEALED DOG CAPSULE

COSMIC-RAY INSTRUMENTS

BURNED-OUT
ROCKET ENGINE

watertight seal around his neck— "the best thing I did all day," he later recalled. He asked the helicopter pilot to give him five minutes to jot down instrument readings. Then he armed the explosive escape hatch.

"I was lying there, minding my own business," he said later, "when I heard a dull thud." The hatch had blown off, the sea was surging in and the next thing Grissom knew he was out of the capsule, thrashing in the water, trying to keep afloat under the down-beating blast from the rotors of three helicopters. His suit was leaking air through the oxygen valve he had forgotten to close and he remembered, in sudden desperation, that he was carrying two heavy rolls of fifty dimes each as souvenirs. Meanwhile, one helicopter managed to get a hook on the foundering capsule. But it was not able to lift the waterlogged weight and, in the end, the cable had to be cut. *Liberty Bell 7* sank in 2,800 fathoms.

Another chopper at length got a horse-collar rescue sling to Grissom, so NASA at least had its astronaut back alive. But losing the capsule hit hard. "I hated to see that capsule finally go down," Grissom said. "In all the time I've been a pilot, that's the first thing I ever lost."

Despite the loss of Grissom's craft, NASA judged the two suborbital flights successful enough to move on quickly to the real thing. Everybody hoped that the projected three-orbit flight would outshine Gagarin's one loop around the world. The hope was short-lived. On August 6, cosmonaut Gherman Titov went aloft in *Vostok II* and stayed up for an astounding 25 hours and 18 minutes—17 orbits around the earth. Even more galling, NASA had to announce that the first attempt to orbit an American would probably slip over into early 1962.

For 40-year-old Marine Lieutenant Colonel John Glenn, the intensive training leading to his flight seemed endless. He logged 59 hours and 45 minutes on the procedures trainer alone, working through 70 simulated missions before he finally lay aboard *Friendship 7,* sitting 10 stories up atop a primed and ready Atlas rocket. About 50,000 "bird watchers" crowded the beaches near Cape Canaveral—some had been camping there for weeks—and another 100 million or so huddled around television sets all over the country. At 9:47 a.m. on February 20, 1962, they witnessed John Glenn's lift-off in a cloud of steam and fiery rocket exhaust.

At 9:52 a.m., the operations team at Mission Control noted in the log: "We are through the gates." The first American astronaut was in orbit.

Fifteen minutes on his way, Glenn saw the Canary Islands and the coast of Africa and the Atlas Mountains coming up. A lavish yellow, orange and purple sunset greeted him over the Indian Ocean, and he flew on through a night that was fitfully illuminated by lightning storms far below. Daybreak arrived 45 minutes later over the Pacific Ocean.

Now the yaw reaction jets began to malfunction, and Glenn had to switch over to fly-by-wire to keep the capsule in the proper attitude for the duration of the mission. Ground stations began relaying enigmatic commands that he check the landing-bag deploy switch. Then Glenn was told to retain the retropack.

Not until he had splashed down was he advised of the problem (although he had deduced it during the flight). One of the black boxes had mistakenly announced that the heat shield was loose. Not knowing for sure, ground control had hoped that the straps on the retropack would hold the shield in place long enough to keep the capsule from incinerating on reentry. Fortunately for the space program, America's first manned orbital flight did not end in disaster.

The nation's reaction, beginning with President John F. Kennedy's congratulatory phone call, was a mixture of pride, relief and consuming joy. "We have a long way to go in this space race," Kennedy said in a speech from the Rose Garden. "But this is the new ocean, and I believe the United States must sail on it and be in a position second to none."

The next three voyages into the new ocean were designed to confirm—and extend—human ability to function in space. Air Force Captain Deke Slayton, 38, had been the one scheduled to make the fourth flight, but at practically the last minute the doctors announced that he had an erratic heartbeat. Slayton was scrubbed from the mission, but soon afterward he was appointed coordinator of astronaut activities.

Thus the next flight went to Navy Lieutenant Scott Carpenter, 37. No flight that brought both astronaut and capsule back intact can properly be called a failure, but the flight of *Aurora 7* was as dicey a mission as the Mercury program would see.

Lift-off went nicely, and once in orbit Carpenter switched to fly-by-wire to flip the capsule over to fly heat shield forward. Throughout the mission Carpenter made enthusiastic use of the fly-by-wire and manual controls to slew the capsule around to take photographs and watch for ground flares. In addition, six times in the first two orbits he accidentally activated both the automatic and the manual systems, depleting fuel in both tanks to below 50 per cent. In order to have enough control fuel for reentry, Carpenter had to spend more than an hour of the last orbit in drifting flight.

When the time came for retrofire, Carpenter was in the manual control mode and somewhat behind on his pre-retrofire check list. Ground control instructed him to switch to the automatic stabilization system, but when he did, he found that it would not hold the capsule in the proper retrofire attitude. He hurriedly switched to the semiautomatic fly-by-wire and brought the capsule into alignment. But in all the switching, Carpenter had forgotten to turn off the manual system, and for 10 long minutes he was spending fuel from both tanks.

Moreover, he had underestimated the capsule's attitude error. The misalignment would be responsible for a 175-mile error on reentry. Then, when he bypassed the misbehaving automatic system and punched the button for the retrorockets, they fired three seconds late—adding another 15 miles to the error. The retrorockets also suffered a loss of thrust, which contributed yet another 60 miles to the overshoot. If Carpenter had not fired the rockets manually, the error would have been even greater.

He hit the water in good shape—but 250 miles beyond where he should have been, and well beyond the range of the communications network. A lengthy period of suspense gave way to a sigh of relief when a patrol plane finally spotted him.

The excessive fuel consumption on *Aurora 7* was worrisome. A principal objective of the fifth Mercury flight, therefore, was for the pilot to refrain from spending fuel in maneuvering the capsule unless he absolutely had to. The job went to Navy Commander Wally Schirra, who was launched in *Sigma 7* on October 3, 1962, for a projected six-orbit flight and the first landing in the Pacific.

Once in orbit, Schirra switched to fly-by-wire briefly to test the controls and edge the capsule over into blunt-end-first flight. Thereafter, for the most part, he switched off all controls, both automatic and manual. "Drifting and dreaming," Schirra called down. He had a little trouble with his spacesuit temperature but, since that had been his specialty, managed to keep it bearable. "It's not worth discussing any more," he reassured the ground.

Advised in the third orbit that he was good for six, he yelled "Hallelujah!" and went on his way. When he splashed down within five miles of his target—the carrier *Kearsarge* north of Midway Island—he still had 80 per cent of his control fuel intact. A ship's whaleboat towed the capsule along-side, and he and *Sigma 7* were hoisted aboard. Analyzing a flawless flight, one engineer back at Cape Canaveral said, "Wally showed us today that we have come a long way."

The Mercury program drew to a close on May 15, 1963, with Air Force Major Gordon Cooper occupying the form-fitting couch in *Faith 7*. The launch proved a repeat of Schirra's unblemished takeoff. "Feels good, buddy," Cooper told Schirra on the ground. "All systems Go."

From orbit, Cooper was surprised at the detail he could make out—the wake of a boat on the Nile, tendrils of smoke from villages in the Himalayas. He was supposed to get some sleep, but the best he could do was to catnap. He was too excited, and besides, his weightless arms floating around kept waking him up.

Trouble of a sort developed in the 19th orbit, when a small light indicated that G forces were building early. Two orbits later the minor glitch became major—the automatic stabilization and control system lost all power. "I figured things must be in quite a sweat down below," Cooper said afterward, "but I was privately pleased. I was glad for this chance to do the job myself and I was certain by now that I really had the thing nailed."

He did. *Faith 7* splashed down four miles away from the *Kearsarge,* 34 hours and 20 minutes after lift-off. The helicopters relayed Cooper's polite request—as an Air Force officer—for permission to board.

Back in 1961, only 20 days after Shepard's first ballistic shot, President Kennedy had declared that the United States should send a man to the moon within the decade. Successful completion of the Mercury program seemed to have moved the nation well along toward that goal. It had been costly—55 months, involving the efforts of 7,300 contractors and subcontractors, more than two million people and more than $400 million. But for the United States, Mercury was only the beginning. □

VON BRAUN'S ROCKETS

A rocket slung over his shoulder, 18-year-old Wernher von Braun (right) heads across a testing ground—a disused firing range—with Rudolf Nebel, a fellow rocket enthusiast.

Von Braun, dressed in a dark suit (right), escorts high German officers—submarine chief Admiral Karl Dönitz walks at left front—on a tour of the rocket center at Peenemünde in 1943.

"For my confirmation," Dr. Wernher von Braun once said, "I didn't get a watch and my first pair of long pants, like most Lutheran boys. I got a telescope." That gift confirmed von Braun's passion for the stars and helped launch a career—more accurately a lifetime—spent trying to reach them with rockets. At the age of 12 he designed a rocket-powered wagon; and in 1933, when he was 21 years old, he made his first sketch for a rocket to the moon.

One year before that, however, von Braun had been discovered by the German Army and became its top civilian specialist on rocketry. Thereafter he had to nurse his dream of outer space in private. (The rocket he designed for the V-2 missile worked perfectly, he told a friend when the first V-2s hit London, except that it landed on the wrong planet.) In 1945 von Braun led hundreds of his top scientists away from the advancing Red Army to surrender to U.S. forces in Bavaria. He and 118 others were hired and sent to Fort Bliss, Texas, to refine the V-2 and explain how it worked.

In 1950, when the Korean War began, the von Braun team moved to Redstone Arsenal in Alabama to build a long-range nuclear missile. A variant of the resulting Redstone carried the United States into the space race. It powered both *Explorer I,* launched in January 1958, and the first U.S. manned flight three years later.

By then, von Braun's engineering skills and visionary leadership had made him a hero to his adopted country. Yet he never lost sight of his goal. Asked what it would take to build a rocket to reach the moon, he answered simply, "The will to do it." Made the head of Redstone's new Marshall Space Flight Center, von Braun put that will to work on the biggest job of his career: building the massive Saturn V rocket for Project Apollo. In July 1969, *Apollo 11* fulfilled not only a U.S. pledge but von Braun's own lifelong dream to put a man on the moon.

His left arm in a cast following an automobile accident, von Braun stands with other top German missile experts after surrendering to the U.S. Seventh Army in Bavaria on May 3, 1945.

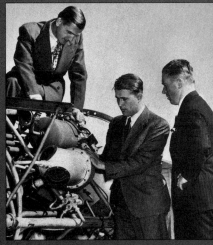

Von Braun (center) and two colleagues, under contract to the U.S. Army, look over the motor of a German V-2 rocket at White Sands, New Mexico, in 1946.

At a ceremony in 1957, Defense Secretary Charles E. Wilson decorates von Braun with the Civilian Service Award—the Defense Department's highest honor for a civilian.

Von Braun *(above, right)* checks recording equipment used during test firing at the Redstone Arsenal near Huntsville, Alabama, soon after the Russians launched *Sputnik I.*

Seated with Dr. Ernst Stuhlinger, head of the Research Projects Office at Redstone, von Braun holds a model similar to *Explorer I,* launched into orbit on the 31st of January, 1958.

Von Braun oversees a technician working inside a Redstone body unit—the sophisticated cap that housed all of the missile's guidance and control instruments. As the head of research and design at Redstone, von Braun was in charge of about 3,500 scientists and technicians.

Wearing safety helmets, von Braun and two engineers confer at the Redstone test site. The spray behind them was used to help cool down the missile stand's exhaust deflector.

Inside the Fab Lab—Redstone's immense fabricating laboratory—von Braun inspects a missile's radar in late 1957. "I've never felt my job was to sit in my office and think," he said. "I want to know what my baby will look like."

Seated in his office, von Braun reads some of his fan mail. He lobbied tirelessly to advance his dream of conquering space. "I have to be a two-headed monster," he once remarked, "scientist and public-relations man."

Von Braun meets with his lab chiefs to discuss problems in the development of the Jupiter missile, a modified Redstone that powered the *Explorer I* satellite.

Taxiing their rented plane down the runway, von Braun gives his wife, Maria, a flying lesson. "Like skin-diving," he said, "flying seems to give me a mastery of the third dimension."

Von Braun peers through a 16½-inch telescope at the Huntsville observatory, built by the local astronomical society. The president of the society: Wernher von Braun.

Von Braun, his wife, Maria, and one of their children enjoy afternoon tea at home. Von Braun lived in Huntsville until 1970, when he took up a NASA post in Washington, D.C.

TRAINING FOR SPACE

To prepare the Mercury astronauts for their space missions, NASA invented an extraordinary array of training procedures and equipment.

Some of the gear was designed simply to keep the astronauts alive beyond earth's atmosphere. Their spacesuits were designed to be self-contained minienvironments in case of emergency. Other equipment was used to subject the men to ordeals that simulated the unearthly conditions they could expect to encounter during their flights. Overdosing on carbon dioxide, for example, and enduring sessions in heat and pressure chambers were two of the less rigorous features of astronaut training.

As practiced test pilots, all seven astronauts were familiar with the sensation of "pulling high Gs"—experiencing the acceleration that slams a jet pilot with several times the force of earth's gravity. Now they were required to take high-G punishment in greater doses than ever before in order to accustom themselves to the pressures they would feel as they powered off the Cape Canaveral launch pad and again as they plunged through the atmosphere on reentry. The machine that meted out this punishment was known as the Wheel, a centrifuge with a 6-by-10-foot gondola at the end of a 50-foot rotating arm powered by a 4,000-horsepower motor. Each astronaut spent hours strapped into the gondola, being swung at speeds that created accelerations as high as 16 Gs—a force at which a 200-pound man, for example, feels as though he weighs one and a half tons.

One particularly grueling Wheel exercise involved flipping the gondola as it swung so that the direction of acceleration changed abruptly. In two seconds the man inside would be hurled through a range of 18 Gs—from being pressed into his seat with a force of 9 Gs ("eyeballs in," the astronauts called it) to being thrown forward against the restraining straps with an equal force ("eyeballs out"). In John Glenn's laconic

In this double-exposure photo, Deke Slayton practices in the ALFA (Air Lubricated Free Axis) trainer, tilting and rotating his couch with compressed-air jets controlled by a mechanism under his right hand, just as the astronauts in orbit would maneuver their capsules by means of hydrogen peroxide jets. A screen located between his feet shows horizon maps of earth, so he can try adjusting capsule attitude in relation to the planet.

Alan Shepard checks the fit of his individually molded couch, used for high-G training as well as during flight. Lining the wall of the ''throne room'' are the contour couches of the other six astronauts.

account, "We were bouncing around a bit but it was quite tolerable."

An even wilder ride, one that pushed each astronaut's vertigo tolerance to the limit, was provided by a beast of a machine called the MASTIF (Multiple Axis Space Test Inertia Facility). The MASTIF had three cages mounted one inside the other on gimbals; a passenger strapped inside the innermost cage could be spun in three directions at once, to simulate the simultaneous rolling, pitching and yawing possible in space. Gus Grissom offered this visceral description of a dizzying spin of 30 revolutions per minute in the MASTIF: "The organs of your body seem to be sloshing in all directions, and your head and feet feel full of blood. You now have only a short time before violent illness overcomes you."

A more pleasant sensation associated with space flight—but one more difficult to simulate—was weightlessness. Astronauts tumbling in the MASTIF or straining against high Gs on the Wheel were typically subjected to greater stresses than they expected to undergo on their actual missions. But the best that could be done to duplicate zero gravity was to take an astronaut up in a jet aircraft for brief interludes of weightlessness. As the jet pilot pulled out of a steep, high-altitude climb, the plane would coast for about 60 seconds across the top of its parabolic flight path while the astronaut, weightless, tried simple things such as eating and drinking. It was at best only partial preparation for the prolonged weightlessness in store for the space traveler.

Accustomed as they were to flying erect in an airplane pilot's seat, the astronauts had to practice on a space-flight simulator to get used to steering the spacecraft lying down. And as insurance against an other-than-normal landing, they also practiced a variety of survival techniques, including those for jungle and desert survival, as well as emergency procedures for escape from a sinking capsule after splashdown. □

Gus Grissom *(left)* and **Wally Schirra** *(below)* feel the pull of 8 Gs as the centrifuge whirls them through a simulated takeoff and reentry. At this speed, even speaking into the microphones in front of them is a strain.

The astronauts try on their spacesuits at one
of innumerable fittings for these custom-made
garments. In Schirra's words, they required
"more alterations than a bridal gown." The long
johns worn under the suit were equipped with
ribbed panels to facilitate the circulation of air.

What the well-dressed space explorer was
wearing in 1960: a suit of rubber and aluminized
nylon topped by a fiberglass helmet with
acrylic visor. The spacesuits were tested during
training in conditions that ranged from
weightlessness to underwater submersion.

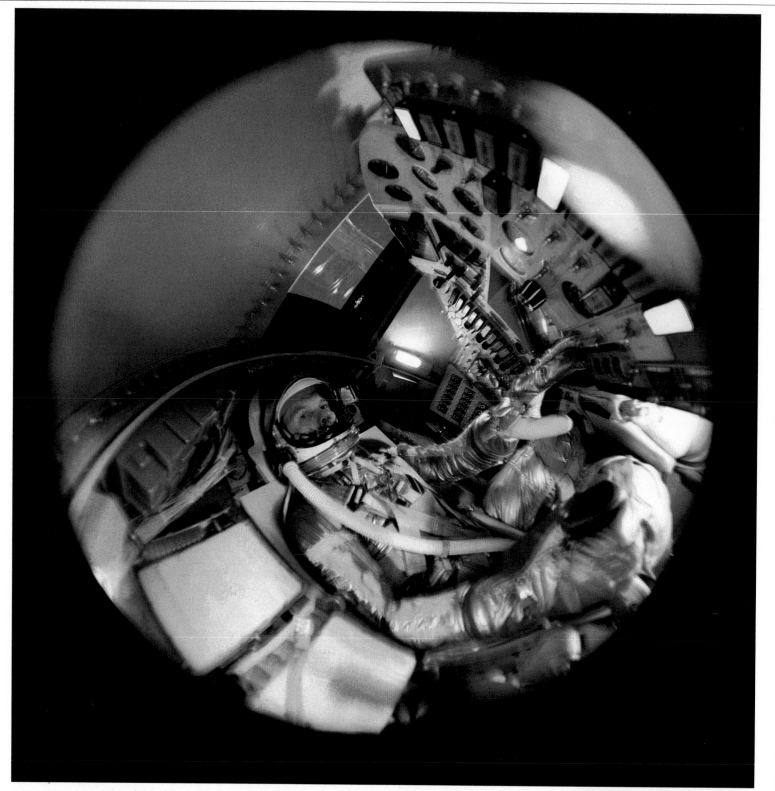

Inside a model space capsule, John Glenn checks out the instrument panel that he helped design for his orbital mission. The capsule window is coated with a special color filter to counteract the blinding sunlight beyond the protection of the atmosphere.

Lines of light trace the disorienting three-way path traveled by an astronaut twisting and spinning inside the MASTIF trainer *(right)*. Practice in halting this machine's wild ride taught an astronaut how to prevent his capsule from tumbling out of control once actually in space.

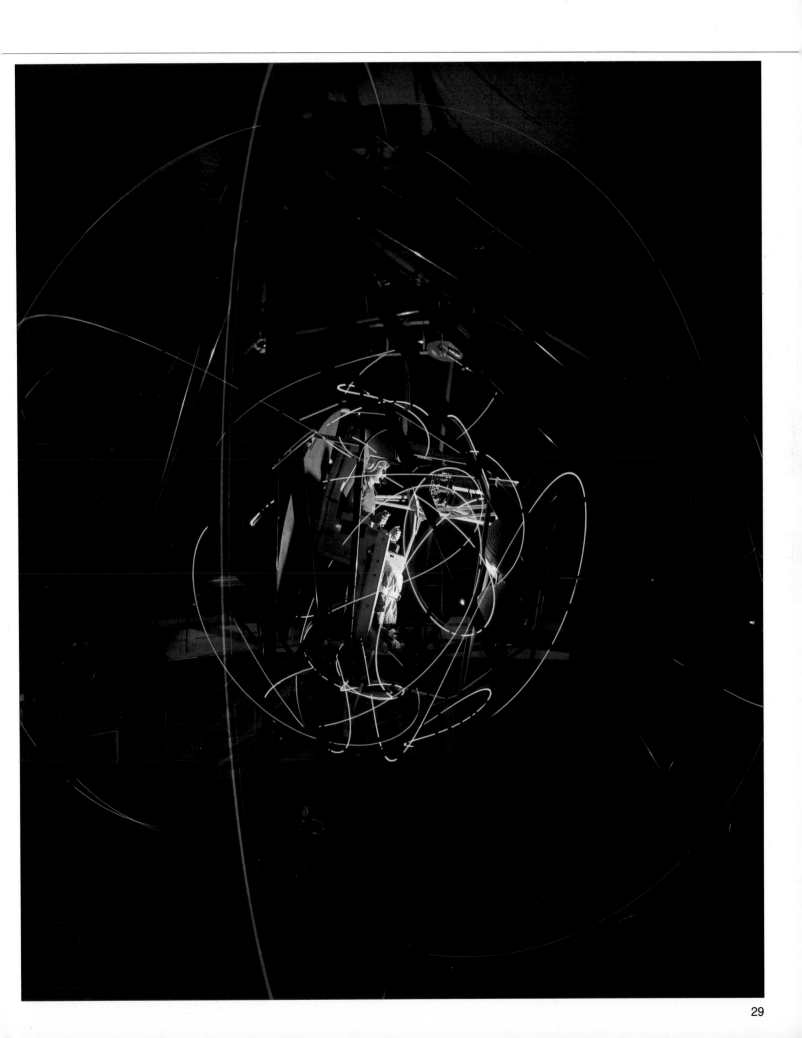

John Glenn shelters from the desert sun under a crude tent made from a parachute. Air Force experts in desert survival put the astronauts through their paces for several days in the sun-parched land near Fallon, Nevada.

On a desert survival exercise, a scruffy band of astronauts display their Arab-style burnooses made from strips of parachute cloth. Some of the men went without water for long periods to test their responses to dehydration.

Supervised by a Navy survival instructor in a large raft, Deke Slayton *(foreground)* consults a chart while behind him astronauts Schirra, Grissom and Shepard flash signal mirrors from their life rafts in the Gulf of Mexico.

Fully suited up, Gus Grissom wriggles out the capsule's narrow neck exit in an escape drill at sea. The exercise tested not only the agility of the astronaut himself but also the buoyancy and watertightness of the spacesuit.

ASTROCHIMPS GO FIRST

Squeezed into Mercury's bottle-shaped spacecraft, John Glenn is visible through a hatch that would be bolted shut before blast-off.

Monkeys Able *(left)* and Baker make the cover of *Life* as America's first live passengers to return safely from space. Able later died when doctors tried to remove a tiny electrode that measured her heartbeat.

A seven-pound rhesus monkey named Sam emerges from his prototype Mercury capsule in December 1959 after a 13-minute ride that took him 55 miles into the upper atmosphere.

Before astronauts there were astro-chimps—and mice, fruit-fly larvae and mold spores. Most of what went up had the simple job of coming down again for study. But monkeys and chimpanzees had a more complex mission. They were the best test of whether humans could survive in space. And they were smart enough for basic mechanical tasks. If they could pull levers while weightless or during fast acceleration, the thinking went, so could an astronaut.

Many young primates were trained, with the 20 chimpanzees undergoing a program almost as rigorous as the astronauts' own. The chimps were tested at the Holloman Aerospace Medical Center in New Mexico. They endured both zero gravity and high G forces. And they learned to manipulate levers at timed intervals, receiving banana-flavored pellets or mild electrical shocks—depending on how well they mastered their tasks.

Four monkeys and two chimps finally flew (one monkey drowned in an unrecovered nose cone). But the chimpanzees proved the effectiveness of the Mercury capsule's life-support systems and performed their lever assignments well. They also gave ground crews a chance to perfect countdown, monitoring and recovery procedures. And by taking it all with aplomb, they helped boost the confidence of astronauts and engineers alike.

Able and Baker "brief" the Mercury astronauts in this cartoon from a New York newspaper.

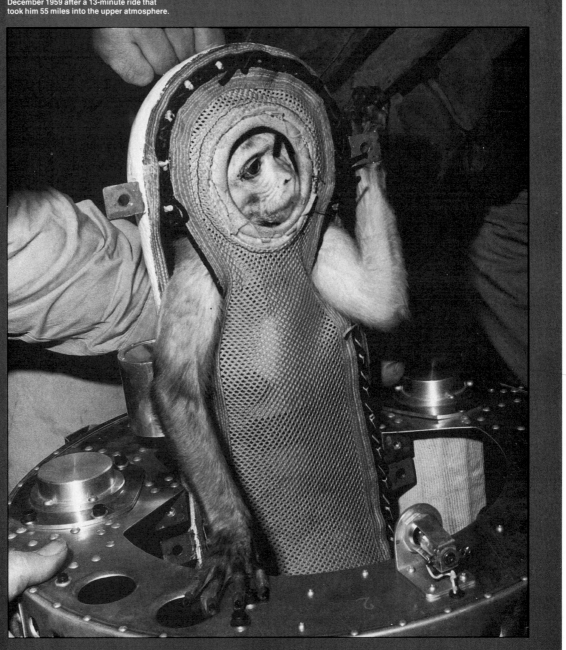

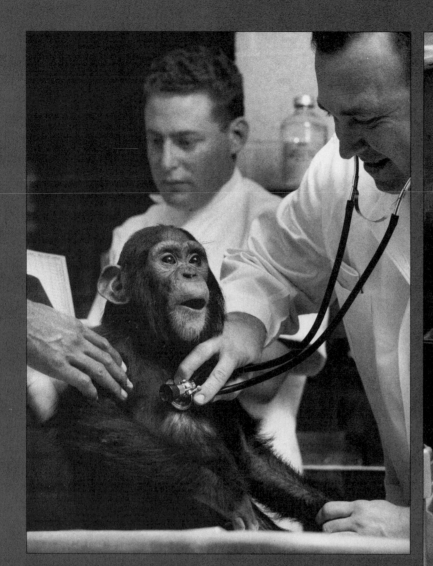

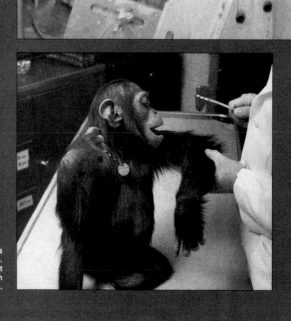

A good-natured chimpanzee named Billy submits to a routine chest examination at the Holloman Aerospace Medical Center. Later he turned the tables—seizing the stethoscope and listening to the doctor's heart.

Billy opens wide for his morning physical checkup. The chimps were fed antibiotics mixed with liquid raspberry gelatin to cure their jungle diseases—and got lemon drops as a reward if they behaved in examinations.

Wearing his regulation chimp tag, Rufe eyes the rubber hammer during a test of his reflexes. The chimps were obtained young so that they would be amenable to training and remain docile throughout their stay at Holloman.

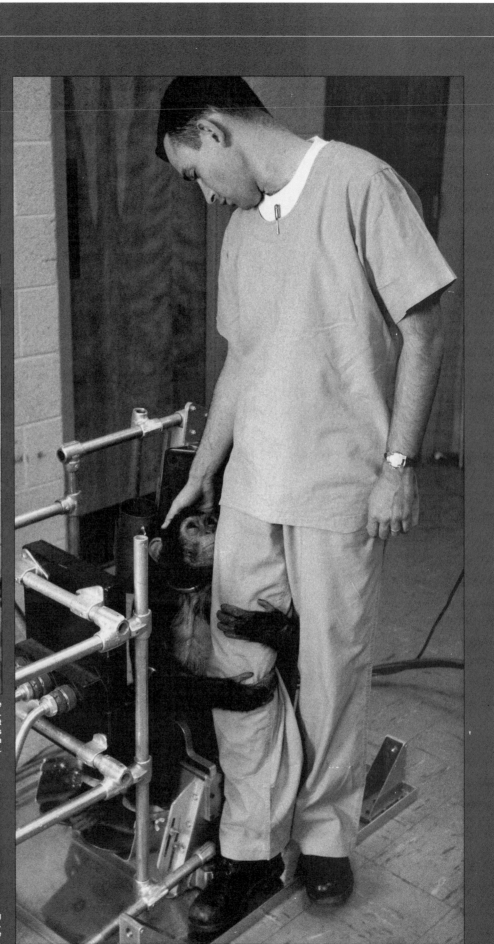

Strapped in at the legs and neck, Bobby Joe works one of the Holloman learning machines. Pulling the correct lever made a reward pellet drop into the tray. One chimp pulled 7,000 levers in 70 minutes—and made fewer errors than a human VIP who tried to match the feat.

After one exercise, Bobby Joe swaps a hug for a scratch from one of his Air Force keepers. Forty-five veterinarians and technicians were assigned to the 20 chimpanzees at Holloman.

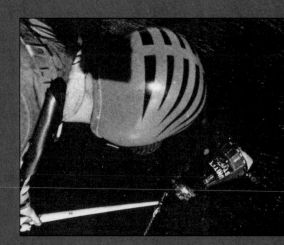

A helicopter crewman snags the Mercury capsule that contains a 37-pound chimpanzee called Ham—he was named for Holloman Aerospace Medical Center—after his suborbital flight on January 31, 1961.

Technicians open the sealed couch in which Ham was confined. Ham—who had 15 months of training for the 17-minute flight—endured 18 Gs and 6.6 minutes of weightlessness.

Savoring a postflight apple, Ham lies in his couch awaiting physical examination. The other half of the couch held the levers that Ham worked almost perfectly throughout the flight.

Ham grins in seeming triumph as doctors prepare to remove the wires that measured his reactions to the flight. A camera mounted inside the capsule recorded Ham's every move.

GAGARIN: SOVIET TRIUMPH

A patriotic song, "How Spacious Is My Country," heralded the radio and loudspeaker announcement in the Soviet Union on April 12, 1961. "The world's first spaceship, *Vostok* ['East'], with a man on board, has been launched. . . . The first space navigator is Soviet citizen pilot Major Yuri Alekseyevich Gagarin."

The historic flight was a double first for the Soviets: Cosmonaut Gagarin did not just leave the earth, he orbited it, then safely parachuted down near the Volga River 25,000 miles and 108 minutes after takeoff. More than 1.1 million pounds of thrust boosted the *Vostok I* to a speed of 17,400 miles per hour and an extreme height of 203 miles above sea level—and that with a spacecraft almost three times as heavy as America's Mercury.

The Soviet reaction, predictably, was a mixture of jubilation and gloating—complete with banner headlines in bold red ink. "You have made yourself immortal," Premier Nikita Khrushchev told Gagarin. "Let the capitalist countries try to catch up." Gagarin's face suddenly appeared on stamps, book jackets and posters everywhere. Gagarin himself—quickly nicknamed Gaga—appeared atop Lenin's tomb in Red Square to review a parade, receive a 20-gun salute and listen to long patriotic speeches in his honor. Khrushchev made the "Space Swallow" a Hero of the Soviet Union, then—sensing something inadequate in that distinguished old title—created a new title: First Hero Cosmonaut.

The American reaction was just as predictable. Associated Press bureaus, polling the Joe Smiths in local phone books, found some admiration for the flight itself and much dismay at having been beaten, once again, by the Russians. That the United States might repeat the flight was scant consolation. "Everyone remembers Lindbergh," cried one Congressman. "But who remembers the second man to fly the Atlantic?"

Cosmonaut Yuri Gagarin of the Soviet Air Force strides the red carpet laid for him at Moscow airport, where Soviet dignitaries waited to honor him for his journey around the world in the spaceship *Vostok I*. The 27-year-old test pilot had been jumped from senior lieutenant to major just before the *Vostok* flight.

Accompanied by former President Kliment Voroshilov *(left)* and Premier Nikita Khrushchev, a smiling Yuri Gagarin waves to the Red Square crowd from the top of Lenin's tomb—the symbolic pinnacle of Soviet society.

Celebrating Muscovites crane their necks for a glimpse of Gagarin during the Red Square rally on April 14, 1961. To greet the cosmonaut the crowd chanted, *"Slava Yuri"*—"Glory to Yuri."

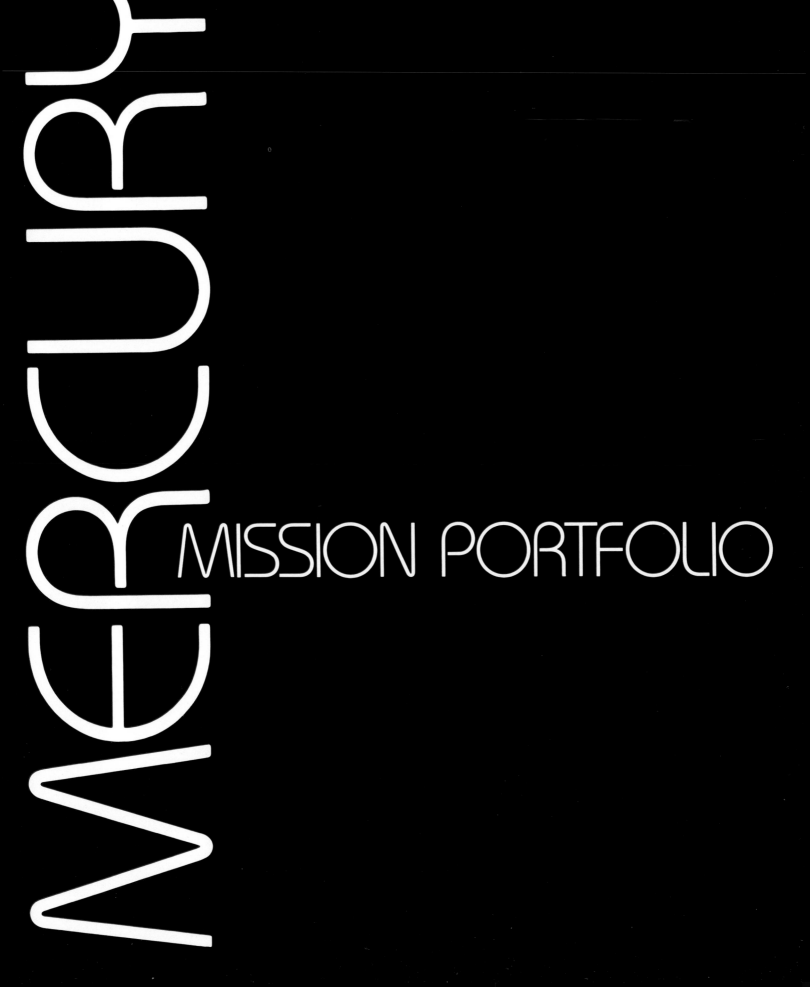

FREEDOM 7

"The butterflies were pretty active," recalled Alan Shepard. The steel-nerved test pilot's momentary anxiety was mild compared with the apprehension felt by others involved in launching *Freedom 7* on May 5, 1961. America's first attempt to shoot a man into space, the 15-minute suborbital mission was a probe exploring a wide range of unknowns, questions that space chimps could not—and Russians would not—answer. Concern centered on the effects of weightlessness; some psychologists even suggested that it could affect the mind. But Shepard allayed all fears, performing normally during the gravity-free state and passing the postflight physical exam easily. He also proved that an astronaut could operate the systems that maneuver the capsule and noted details that would make future manned flights easier and more predictable.

Outfitted in a silvery pressure suit and helmet, Shepard leaves the living quarters at Cape Canaveral just before 4 a.m. for the trip to the launch site. Astronaut Gus Grissom follows.

Carrying a portable air conditioner for his suit, Shepard steps from the transfer van at 5:15 to enter the gantry elevator. A graying sky and a computer glitch delayed lift-off until 9:34.

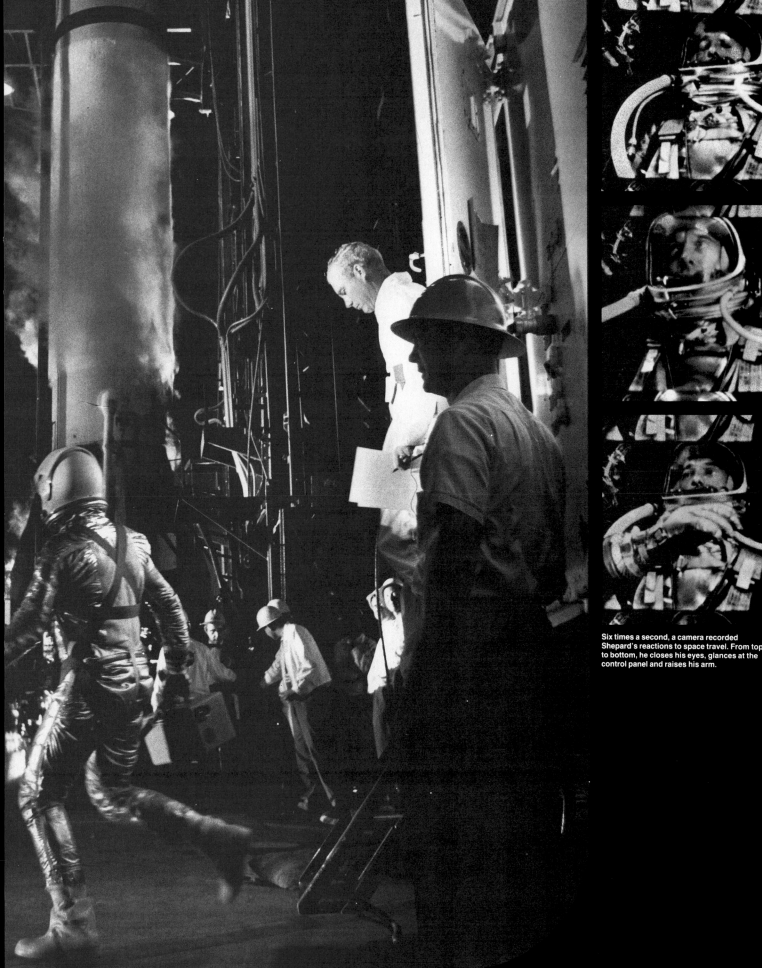

Six times a second, a camera recorded Shepard's reactions to space travel. From top to bottom, he closes his eyes, glances at the control panel and raises his arm.

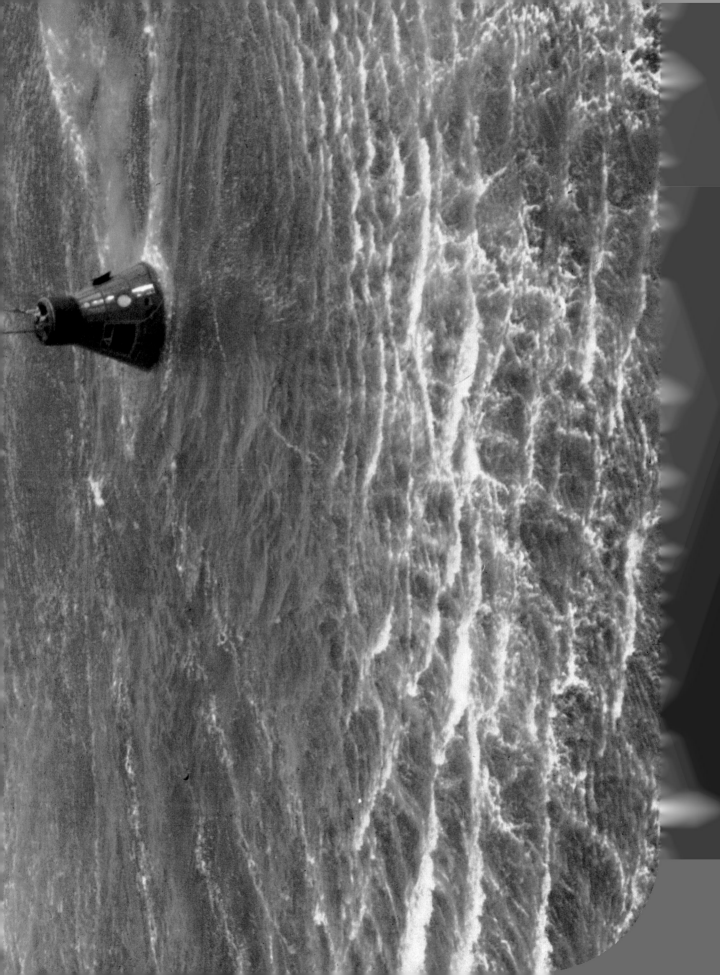

...lf hour after lift-off, the space pioneer ...triumphantly aboard the carrier *Lake ...mplain* after the pickup helicopter has ...ed his capsule on deck. In a thorough ...fing, Shepard received a full physical ...nd dictated a lengthy, detailed report.

...House three days later, President ...s NASA's Distinguished Service ...epard. Afterward, crowds cheered as ...e Capitol for lunch.

LIBERTY BELL 7

The flight of *Liberty Bell 7* on July 21, 1961, essentially repeated that of *Freedom 7,* but the capsule was closer in design to the world-circling crafts to come, boasting a large windshield-like window instead of two meager side ports. To determine an orbiting astronaut's ability to navigate by earth land-marks, Gus Grissom spent much of his 15 minutes aloft peering out. In fact, he was so awed by the 800-mile-wide panorama and his ability to spot familiar sites that he did not completely check out the ship's manual controls. But he did demonstrate that the new rate-stabilization controls—rather like power steering in a car—made it easier to control the craft's pitch, roll and yaw. Other improvements included static-reducing microphones in the helmet, an easier-to-scan instrument panel and an easier-to-open hatch. The old hatch, with its heavy locking mechanism, gave way to one that was blown off by tiny bolt-shearing explosive charges. A malfunction in those charges produced the tense moments shown in the pictures here. After a safe splashdown, the bolts detonated early. Grissom escaped, but *Liberty Bell 7* shipped water and sank.

As *Liberty Bell 7* sinks almost out of sight, Grissom swims back *(left)* to attempt to help the helicopter snag it with the recovery cable, then watches as the capsule rises from the water *(above).* But the waterlogged load was too much for the chopper's engine; the cable was cut and *Liberty Bell 7* sank into the sea.

Grissom dangles awkwardly as another chopper hoists him on board. In the close-call rescue, the astronaut himself was sinking as his suit leaked air through an open valve.

Aboard the carrier *Randolph*, Grissom starts to take off his pressure suit. The dual strains of piloting through space and nearly drowning had left him quite exhausted.

TITOV: "I AM EAGLE"

Cosmonauts Yuri Gagarin *(left)* and Gherman Titov embrace at the Moscow airport after Titov's 17-orbit voyage aboard *Vostok II*. The 26-year-old Titov had been Gagarin's backup pilot for *Vostok I* four months earlier.

This photodiagram follows Titov and *Vostok II*, orbit by orbit, on their spiral route through space. As the space capsule circled the earth, the earth itself was rotating beneath.

"I am Eagle, I am Eagle," crackled the voice from outer space, showering its greetings on all below. "I wish you had it so good." Thus spoke Cosmonaut Gherman Titov, code name Eagle, who on August 6, 1961, became Russia's second and the world's fourth space traveler. Titov orbited the earth 17 times in 25 hours and 18 minutes: the first multiple orbit and the longest single journey—nearly 435,000 miles—ever made by a man.

For all that, it was a rather lazy day aloft. Titov tested *Vostok II's* manual controls on the first orbit, had lunch on the third—a three-course meal squeezed out of tubes—and supper on the sixth. He slept away five more orbits, indeed overslept his scheduled seven and a half hours by 35 minutes. ("I am turning in now," he said. "You do as you please, but I am turning in.") He awoke on the twelfth orbit, ate breakfast on the fifteenth, then parachuted down two orbits later into a field 460 miles southeast of Moscow.

The Soviets were uncharacteristically open about the flight. Russian journalists were allowed to witness the launch of *Vostok II*—not the case with *Vostok I*—and Titov's broadcasting frequency was announced to the world. Premier Nikita Khrushchev promoted Titov from captain to major and from candidate to full party member in midflight. Cosmonaut Yuri Gagarin flew back from a Canadian tour to greet the new hero, who had proved that humans could endure prolonged weightlessness and who—for all his banter aloft—acknowledged that he had experienced "a homesickness for the earth."

A crowd parading with flags, banners and portraits —including one at left of Khrushchev holding the dove of peace—welcomes Gherman Titov to Moscow's Red Square.

FRIENDSHIP 7

In the eyes of scientists and the public alike, the two suborbital manned Mercury flights and an orbital flight by a chimpanzee named Enos were merely rehearsals. The real drama began on February 20, 1962, when a powerful Atlas missile propelled John Glenn and his craft, *Friendship 7*, skyward in an attempt to achieve the first earth orbit by an American astronaut. The mission's prime objective was to complete three orbits and bring the astronaut back safely. But Glenn also had a full list of experimental tasks. He proved, for example, to be an accurate judge of distance in space as he watched a jettisoned booster drift away. His performance of a series of head and body exercises eased fears that Cosmonaut Titov's reported dizziness was an endemic space disease. In daylight, Glenn spotted coastlines and by night, an Australian city's lights. Other planned tasks were curtailed when the automatic controls went awry as the first orbit ended. But by taking charge of the craft, Glenn made a strong case for the need to have an astronaut aboard. Most important, Glenn's successful mission answered basic questions about the spaceworthiness of the capsule and the ability of a man to endure prolonged weightlessness. It also buoyed a national pride punctured by the Russians' spectacular achievements. A presidential welcome home, an address to Congress and extravagant motorcades marked an emotional reaction not seen since Lindbergh returned from his solo transatlantic flight 35 years before.

As she watches the lift-off on television at home in Arlington, Virginia, Annie Glenn's face reveals controlled apprehension about her husband's fate. Sitting behind her are her two children and the family minister.

Belching enormous clouds of fire and steam, the towering Atlas rocket strains to rise from the pad. Inside the capsule, Glenn's pulse rate rose only to 110 beats per minute, compared with

...smerized by the television screen, Annie
...enn watches the rapidly disappearing tail of
... rocket thrusting her husband into orbit.
...alf hour later, backup pilot Scott Carpenter
...ed to assure her that all was going well.

Lyn Glenn, 14, keeps a stoic vigil while a pre-
space-age hourglass tracks the time. Some of
her father's glory may have rubbed off on
the ninth grader—she was later overwhelmingly
...elected as president of her class.

Annie's relief that her husband's capsule has been sighted at splashdown turns to elation when she hears that he has been retrieved from the Atlantic by the destroyer *Noa*, the ship that won the race to get to him first.

Wearing his earth-side clothes again, Glenn rides on a hoist to the helicopter that will carry him from the recovery ship to the carrier *Randolph*, where officials and physicians will complete the postflight debriefing.

Reunited after the flight, John and Annie Glenn exchange a kiss. The astronaut then surprised his wife with a special gold-and-ruby lapel pin that he had carried with him on his successful journey into space.

Back at Cape Canaveral 72 hours after lift-off
an exuberant Glenn gives a thumbs-up sign as
he shares an open car with President John F.
Kennedy. After receiving the NASA Distinguished
Service Medal, the astronaut took the Chief
Executive for a personal look at the capsule.

A record-breaking blizzard of ticker tape—nearly 3.5 tons—inundates Glenn during a tumultuous welcome in lower Manhattan. An estimated four million New Yorkers braved blustery weather to honor the space hero.

Escorted by Vice President Lyndon Baines Johnson, the Glenns acknowledge the cheers of a quarter million Washingtonians. Later the astronaut addressed a joint session of Congress—an honor that is ordinarily reserved for visiting heads of state.

At one of his New York City stops, Glenn waits for the cheers to subside before making a few remarks. While in the city, Glenn also received honors from the mayor and was given a reception at the United Nations.

Flanked by lines of highway patrolmen,
Glenn receives the more restrained greeting of
his Midwestern friends and neighbors on his
return to his hometown of New Concord, Ohio.
The town's normal population of 2,000
swelled to near 50,000 for the occasion.

When Scott Carpenter's *Aurora 7* lifted off on May 24, 1962, for America's second manned orbital tour, there was still much to learn about the mechanics of space flight. Carpenter delighted in his freedom to test the craft's maneuverability, even flying with his head pointing down to earth—an experience he found not at all disorienting. But the main emphasis was on performing scientific observations and experiments. Carpenter spotted and counted stars, photographed atmospheric and topological phenomena, and recorded how liquid in a sealed flask behaves at zero gravity. Another task—to launch a 30-inch balloon to measure the drag from the ultrathin air—was scrapped when the balloon failed to inflate fully. Other problems included overheating in the pressure suit and cabin, excessive fuel consumption and—most serious of all—a malfunction in the automatic control system that caused a misaligned reentry. A suspenseful nation held its breath for an hour until *Aurora 7* was found—250 miles off course—by the recovery team.

Rising on a column of fire, an Atlas missile thrusts a second American toward orbit around the earth. With only three 15-minute holds, the countdown and the launching went more smoothly than ever before on a manned mission. Even a worrisome morning ground fog began to burn off just in time.

Rene Carpenter exhibits signs of strain *(right)* during the hour after *Aurora 7's* reentry when there is no radio contact and her husband is feared lost. But her mood changes to unrestrained joy *(below)* when she and her son learn that a plane had spotted "a life raft with a gentleman by the name of Carpenter in it."

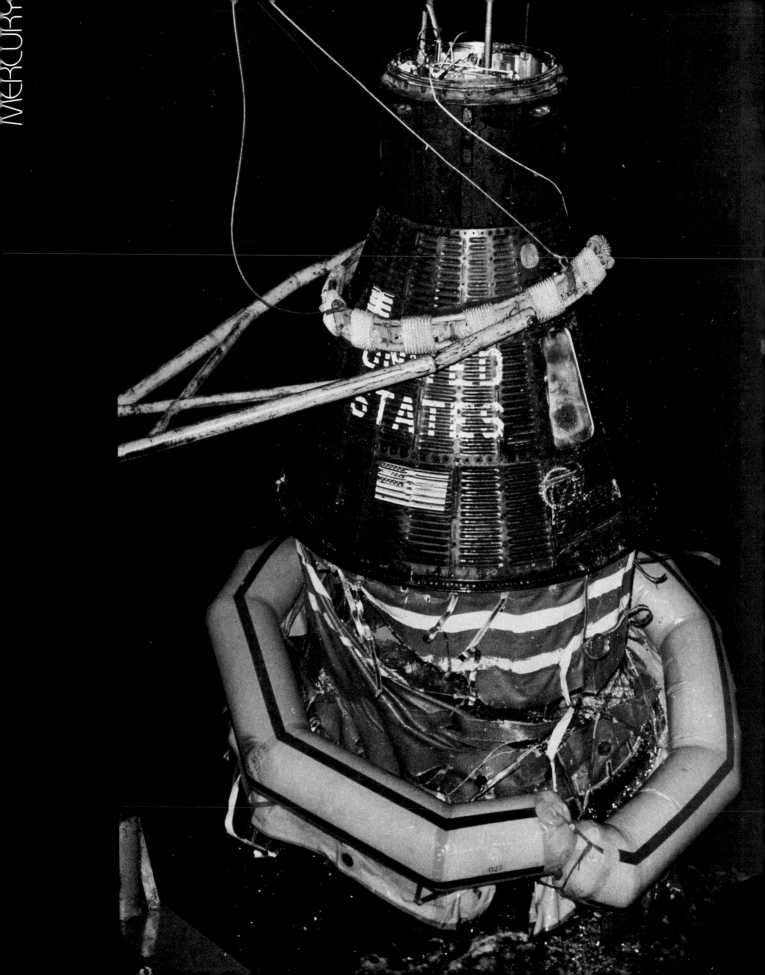

er being fished out of the Atlantic six
urs after splashdown, *Aurora 7* comes to rest
the deck of the destroyer *Pierce.* When
ked up, the spacecraft was listing badly and
taken on about 65 gallons of sea water.

uring the debriefing aboard the carrier *Intrepid,*
Carpenter walks a rail with arms folded to make
the feat more difficult. Navy medics wanted to
determine the effect of four and a half hours of
weightlessness on his sense of balance.

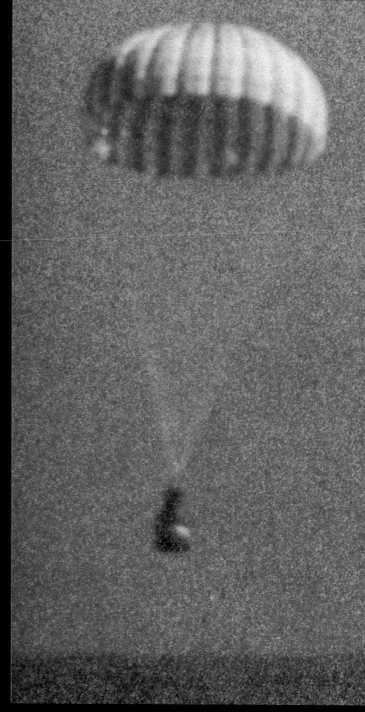

In the van on his way to board *Sigma 7* on October 3, 1962, Wally Schirra dozed off. As the fifth astronaut to make this trip, he was reassured by the Mercury Project's fine record in getting astronauts up and back safely. Moreover, especially compared with Scott Carpenter, Schirra had the advantage of facing only a limited list of chores on his flight: Except for keeping an eye out for ground signals and taking photographs as he saw fit, he had few scientific duties. Instead, his primary job was to curb his piloting instincts and do as little as necessary—to fly in what he tongue-in-cheek called the "chimp configuration." Schirra drastically limited his use of controls, cutting off his power and drifting for long periods during his six orbits. Like Carpenter, he spent much of his time trying to deal with a heat-control problem in his suit. *Sigma 7's* restrained mission was an overwhelming success. Just before reentry, Schirra still had 80 per cent of his fuel—opening the way for a much longer mission.

Dangling from its orange-and-white, 63-foot chute, *Sigma 7* settles down into the Pacific with pin-point precision—only 9,000 yards from the recovery carrier—permitting the event to be captured on film for the first time.

As frogmen finish putting a flotation collar around *Sigma 7*, crewmen aboard a whaleboat attach a tow line. Schirra had requested the boat because he wanted to be picked up while he was still in the capsule.

Only 39 minutes after splashdown, *Sigma 7* is hoisted aboard the carrier *Kearsarge.* Then the pilot on what was labeled a textbook flight crawled out of the capsule and patted it, saying. ''This is a sweet little bird.''

Sitting in the ship's sick bay in his spacesuit, Schirra receives the by-now traditional congratulatory call from the President. Then the *Kearsarge* delivered him to a tumultuous *aloha* in Hawaii before he flew back to Houston.

The original dream of Mercury planners was to achieve an 18-orbit manned flight. By the time Gordon Cooper was launched on May 15, 1963, the goal had grown to 22 orbits—34 hours in space. To test the effects on the human body of such extended weightlessness, Cooper dutifully ate the prescribed foods, pulled on a rubber exerciser, made periodic checks of his temperature and blood pressure, and collected urine samples. He even slept in space, albeit in fitful naps rather than the full scheduled 7½-hour rest. Weight restrictions severely limited other experiments. But Cooper, an amateur photographer, was well equipped with cameras to take both still and moving pictures. Even with the electrical failure that marred the mission's final orbits, it was, all told, a near-perfect end to the opening chapter of America's space saga.

For the fourth and last time, an Atlas missile lifts off to carry an American astronaut into orbit. Project Gemini, the subsequent stage in the burgeoning space program, would use 472,000-pound-thrust Titan missiles.

On board the recovery carrier *Kearsarge* scarcely 45 minutes after splashdown, a sweat-drenched Cooper peers happily out of his capsule. Medics spent eight minutes examining him in the capsule before helping him out.

Navy swimmers ride the flotation collar as
counterbalances while *Faith 7* is towed toward
the recovery ship. Although Cooper brought
the capsule down using the manual controls, his
bull's-eye landing matched Schirra's.

After a moment's dizziness on stepping out
of the capsule, Cooper is helped by doctors as
he heads for the ship's sick bay along a red
carpet. Hot and dehydrated, he had lost several
pounds since suiting up 37½ hours before.

An admiring teenager slips through police lines to shake Cooper's hand during the motorcade through Washington. Like Glenn, Cooper addressed a joint session of Congress and had a New York ticker-tape parade.

With a delighted press corps looking on, First Lady Jacqueline Kennedy presents her bashful son, John Jr., then two and a half years old, to Cooper on the steps of the White House.

No astronaut enjoyed the seemingly endless prelaunch wait in the spacecraft. "What you'd like to be able to do," Gus Grissom said, "is kick the tires and go." With that wishful metaphor, Grissom was describing an ideal unattainable even when he was a young jet-aircraft test pilot; it was certainly unattainable when it came to the enormous undertaking that was space flight. But with Project Gemini, the two-man successor to Mercury, NASA came close.

Reaching that state of efficiency and routine was far from easy. By the spring of 1965, when Gemini finally got space crews off the ground, it was 18 months behind schedule, and no American had been in space for nearly two years. The Soviet Union seemed well ahead in the race to the moon. Soviet cosmonauts had amassed 507 hours in space, compared with only 54 hours for U.S. astronauts. To top it off, only five days before the launch of the first manned Gemini craft, a cosmonaut took the first walk in space *(page 77)*.

But once launched, Gemini soon redeemed itself—and the U.S. space program. In a period of less than 20 months, it reeled off a spectacular total of 10 manned flights, most of them requiring the nearly simultaneous launch of two vehicles.

Gemini's success was all the more remarkable in that the entire project was something of an afterthought. When Gemini was formally approved by NASA in December 1961, Apollo—the ambitious project that would succeed it—already had been officially under way for nearly 18 months, part of a comparatively leisurely NASA schedule that would culminate in a lunar landing sometime in the distant 1970s. But President Kennedy's dramatic pledge in May 1961 to put Americans on the moon before the end of the decade changed all that, forcing NASA to reconsider its plans for putting men on the moon and bringing them safely home.

Until then, the favored plan was a seemingly simple one known as direct ascent. As the name implied, a spacecraft would be boosted directly from earth to land on the moon, then lift off the lunar surface for the return home by firing its own big, on-board rockets. Such a craft, it was calculated, would have to weigh nearly 75 tons and would require a launch vehicle that generated up to 12 million pounds of thrust. But the proposed booster for the mission, the Nova, was still just a concept—nowhere near being ready before the end of the 1960s.

Thus the planners were forced to look hard at another approach—rendezvous—first proposed by early writers of science fiction. One version of rendezvous—EOR, or earth-orbit rendezvous—called for two launch vehicles, one carrying a spacecraft, the other extra fuel, to meet in earth orbit. Fuel would be transferred to the spacecraft, which would then fly directly to the moon, land, lift off and return to earth. This approach was fuel-costly and, like direct ascent, required the spacecraft to be capable not only of landing on and lifting off from the lunar surface, but of reentering the earth's atmosphere.

Another version of rendezvous, however, solved many of the problems inherent in the other two methods. With LOR, or lunar-orbit rendezvous, two attached spacecraft—with a total weight of only 42 tons—could be launched into orbit around the moon, using only one launch vehicle and about half the booster power required for direct ascent. The smaller vehicle could detach itself and descend to the surface of the moon, then use a small rocket to blast off and rejoin the orbiting mother ship. The lunar lander would then be jettisoned, and the greatly light-ened spacecraft would return home.

Lunar-orbit rendezvous looked like the only practical way to fulfill President Kennedy's pledge. But rendezvous and docking in space were complex—and completely untested—maneuvers. Would they work?

Enter Project Gemini, a program a small group of NASA engineers had been pushing as a way to improve the experimental machine that was the Mercury spacecraft. As a prelude to Apollo, however, Project Gemini would also have to answer at least three other questions crucial to lunar exploration: Could space explorers survive and function during the week or more of weightlessness necessary to get to the moon and back? Could they maneuver and perform useful work in the hostile environment outside the womblike confines of their spacecraft? Could they control with precision their path of reentry back into the earth's atmosphere?

Though it borrowed in the beginning from Mercury technology, Gemini quickly evolved into a true second-generation spacecraft, resembling the Mercury capsule mainly in its familiar bell shape. It weighed nearly twice as much and was about one foot longer and one foot wider to accommodate two pilots, who would be seated side by side.

One key difference was the relative simplicity of the Gemini spacecraft (page 79). In Mercury, systems were crammed into every nook and cranny of the cockpit; the capsule was a nightmare to build, test and repair. Gus Grissom recalled one time when "we actually spent three days just trying to get one nut onto one bolt." In Gemini, however, each system was housed in a separate package on the outside walls of the cabin, permitting what was called a quick fix. If a system malfunctioned during prelaunch preparations, the old module could simply be unplugged and replaced.

Other systems were contained in the nine-foot-long adapter section, which was attached to the spacecraft aft of the astronauts' couches and would be jettisoned before reentry. Here, for example, were the 16 small rocket thrusters that would enable the astronauts to control the attitude of the spacecraft and to change orbits for rendezvous and docking with a target vehicle. (The 26-foot-long Agena target vehicle was the upper stage of an Atlas booster that would be launched separately.)

Another important difference from Mercury was that Gemini's design relied on pilot control rather than merely allowing it. This not only delighted the astronauts—who had chafed at playing backup to the automatic black boxes that had controlled the flight of Mercury—but also reduced the number of redundant systems.

Even the Titan II, Gemini's launch vehicle, made things simpler and more reliable. Where Mercury's Atlas booster used liquid oxygen—which required supercold storage and handling facilities—Titan II used propellants that could be stored at normal temperatures, thus simplifying prelaunch procedures. Moreover, the propellants were hypergolic—fuel and oxidizer burned spontane-

Veteran Gus Grissom *(right)* **and rookie John Young—the first space twins—peer from a model of the Gemini cockpit. Of nine pilots selected to join the astronaut program in 1962, Young was the first to be chosen for a flight.**

ously on contact—eliminating the need for a complicated, error-prone ignition system. The propellants also reacted less violently with each other than did those used in Atlas, making possible another Gemini innovation.

Mercury had been encumbered at launch with a heavy escape tower and a complex sequencing system to free the capsule automatically if the Atlas booster exploded. This was replaced by a simple seat-ejection system like that used in jet aircraft. From launch up to 15,000 feet, the pilot could pull the so-called chicken ring, ejecting the two parachute-equipped seats through the hatches.

While the spacecraft took shape, two new groups of astronauts joined the original seven to help fly it: nine in September 1962, and 14 more in October 1963. As with Mercury, each member of the corps had to master a different phase of Gemini and of the embryonic Apollo program.

Gus Grissom, who had been assigned to Gemini, spent so much time at the McDonnell Aircraft Corporation plant in St. Louis that the craft soon was dubbed the Gusmobile. The cockpits of the first three were specifically designed for his short, stocky frame, and later spacecraft had to be modified to accommodate taller astronauts. Thus, no one was surprised when Grissom was chosen to command *Gemini 3*—the first manned flight after two unmanned missions to test booster and spacecraft. Grissom's pilot was rookie John Young, who was also compactly built and fit comfortably in the Gusmobile.

Both men had a quiet, wry sense of humor that sustained them through nine months of intense preparation. For example, Grissom, whose Mercury capsule *Liberty Bell 7* had sunk when the hatch blew prematurely, dubbed their *Gemini 3* spacecraft *Molly Brown*—for the heroine of the Broadway musical *The Unsinkable Molly Brown.* (NASA officials thought the name lacking in dignity but grudgingly gave in when Grissom

said his second choice was *Titanic.*)

Molly Brown rocketed aloft on Tuesday, March 23, 1965, and close to the end of the first orbit it was Young's turn for a bit of fun. Knowing how much Grissom hated Gemini's dehydrated food, Young casually asked, "You care for a corned-beef sandwich, skipper?" and pulled the smuggled item out of his spacesuit. Grissom took a bite, then stowed the sandwich away lest weightless crumbs floating around the cabin foul delicate systems.

After that, their three-orbit mission was all business, with Grissom firing the rocket thrusters to change orbits and make other maneuvers to demonstrate the capability for rendezvous in space. Only at splashdown in the Atlantic did things threaten to go awry. Grissom momentarily forgot to release the parachute, which dragged the craft underwater—suggesting that *Molly Brown* might not be unsinkable after all.

Grissom, Young and *Molly Brown* had proved that Gemini would fly. Now it was time to be more ambitious. A little more than two months later, on the 3rd of June, *Gemini 4* was launched on the first of the long-duration flights. James McDivitt and Edward White stayed aloft for 97 hours and 56 minutes—twice as long as the total time of the eight U.S. astronauts who had preceded them into space. After splashdown, the two were tired from lack of sleep—the radio and the thump of the thrusters firing kept waking them up—but otherwise in such good shape that White did a little jig when he stepped aboard the recovery carrier *Wasp.*

But what electrified the world was the cosmic ballet White had done on the first day. More than 100 miles above the Pacific, anchored only by the 24-foot-long umbilical that fed him oxygen, he had climbed out of the craft and cavorted in space, using a hand-held jet gun to control his movements. White spent 20 minutes in EVA, or extravehicular activity—

twice as long as the Soviet pioneer space walker Alexei Leonov—and enjoyed it so much he had to be ordered back in.

However spectacular, *Gemini 4's* four days in space still amounted to only half the minimum duration of a mission to the moon and back. "Eight days or bust" was the objective—and the crew's motto—when *Gemini 5* soared skyward on August 21. Command pilot Gordon Cooper had wanted to sew that slogan on the shoulder patch he had designed—a covered wagon like those of earlier American explorers. But NASA balked—fearing embarrassment if the mission did, in fact, go bust.

On the very first day of the flight, NASA's caution seemed justified when the fuel cells—being tried in space for the first time—began acting up. These devices, which convert hydrogen and oxygen into electricity, were the craft's principal power source. Between the seats, the irrepressible Charles (Pete) Conrad sketched the predicament: a covered wagon halfway over a cliff.

On the fifth orbit of this cliff-hanger, veteran flight director Christopher Kraft *(page 84),* at Houston's new Mission Control Center, faced the crucial decision: abort, or gamble that the fuel cells, even with problems, would continue to function.

Kraft gambled. Despite the balky fuel cells, *Gemini 5* got its eight days in orbit, almost doubling the endurance record set by Valery Bykovsky of the Soviet Union two years earlier.

The techniques for rendezvous with another object in space still had to be mastered. The fuel-cell problem had prevented *Gemini 5* from practicing rendezvous with a small radar pod released from the spacecraft. And before that, *Gemini 4* had unsuccessfully chased the orbiting upper stage of its Titan II booster, getting a quick lesson in the paradox of orbital mechanics: Speeding up the spacecraft also moved it into a higher orbit than the target—actually slowing it in relation to its quarry.

A further setback to rendezvous came on October 25. With *Gemini 6* astronauts Wally Schirra and Thomas Stafford aboard and ready to launch on a mission calling for both rendezvous and docking, the Agena target vehicle did not go into orbit.

This mishap prompted a spectacular schedule change. *Gemini 6's* launch was postponed; instead of an Agena, the target vehicle would be *Gemini 7*, the next scheduled flight and a long-duration mission. On December 4, *Gemini 7*—the first half of a space double-header that journalists dubbed "The Spirit of 76"—went into orbit with Frank Borman and James Lovell ready to spend two weeks aloft. Immediately, the launch pad was cleared and preparations were begun for the second half, the rescheduled *Gemini 6*. Eight days later, Schirra and Stafford, atop the Titan, heard the engines roar with ignition and saw the mission clock start, indicating lift-off. But 1.2 seconds later, the engines shut down.

According to mission rules, the command pilot, Schirra, should have immediately pulled the chicken ring and ejected both men. If the booster had climbed even a few inches, the engine shutdown would have brought 150 tons of volatile fuel crashing down on the pad, creating a huge fireball. But Schirra had not felt any movement. Knowing that ejection could result in physical injury—and would surely mean a long postponement of the mission—he chose to trust his senses and tough it out.

The booster had not moved. The clock had started because an electrical plug had prematurely dropped out of the bottom of the booster. Meanwhile the malfunction-detection system had shut down the engines before lift-off—and even before the plug had dropped out—because it had sensed something else wrong. The something—found after an all-night hunt—was a plastic dust cover accidentally left in the booster months before when the gas generator had been removed for cleaning.

Schirra's icy calm paid off. Just three days later—with Schirra urging, "For the third time, go!"—*Gemini 6* finally got off the pad. Schirra and Stafford began their chase 1,245 miles behind *Gemini 7*, which already had been in space for 11 days. But by maneuvering into an orbit slightly lower—and faster—than his quarry, then firing his thrusters to kick up and match their orbit, Schirra caught up with his colleagues in less than six hours. The lack of docking adapters prevented actual docking, but for nearly three orbits the two spacecraft flew in formation—sometimes less than one foot apart.

Having achieved the first rendezvous in space, Schirra and Stafford returned to earth. But *Gemini 7* had nearly three days to go. Borman and Lovell, wearing only long underwear for comfort, did the last of their 20 experiments (including monitoring of their brain waves during sleep) and got bored enough for Lovell to muse about the uselessness of legs in zero gravity. A legless astronaut would be perfect, he said, "because you could

utilize that space for something else." Then they donned their spacesuits and splashed down in time to be home for Christmas.

The main goal for the new year, 1966, was docking with the Agena, which had been modified to eliminate glitches. On March 16, Neil Armstrong and David Scott aboard *Gemini 8* did just that, nosing their craft's one-foot-long docking bar gently into the *Agena 8's* notched collar.

But 27 minutes after this flawless link-up, a wild ride began. The two mated vehicles started gyrating violently, spinning and tumbling end over end. Thinking the Agena was to blame, Armstrong undocked. But the problem was in the spacecraft: Free of the Agena, it rolled even faster, making one revolution per second.

Armstrong was dangerously dizzy, his vision blurring; in desperation he fired the thrusters that position the spacecraft for reentry. The spinning stopped, but use of the reentry thrusters dictated an immediate end to the mission—otherwise a leak might deplete the remaining fuel and prevent the return home. *Gemini 8* was brought down safely in an emergency landing area in the Pacific.

Trouble also plagued *Gemini 9*. The prime crew, Elliot See and Charles Bassett, died when their T-38 jet crashed in St. Louis next to the building where their spacecraft was being assembled. Their replacements, Tom Stafford and Eugene Cernan, had one problem after another. The launch was scrubbed twice during countdown—once because the Agena they were supposed to dock with nose-dived into the Atlantic. A substitute target, the ATDA, or Augmented Target Docking Adapter, was sent up. But when Stafford and Cernan finally launched on June 3, they found that the shroud covering the target's nose had failed to jettison. Docking was impossible.

Worse, on the third and last day of the flight, Cernan left the capsule for a scheduled 167-minute space walk

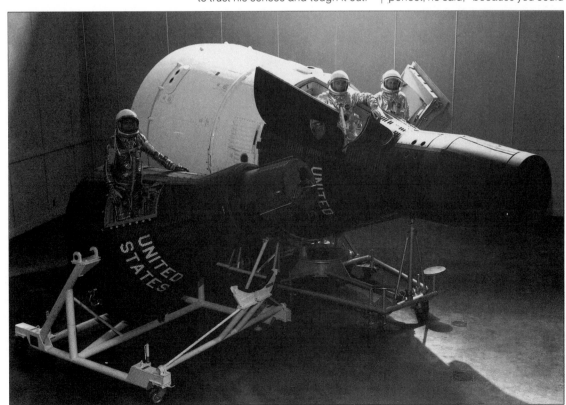

Looking, according to Gus Grissom, "like a Mercury capsule that threw the diet rules away," the Gemini spacecraft overwhelms its smaller predecessor. Gemini's elongated nose held radar gear, reentry thrusters and a parachute.

and ran into problems. From the adapter section he was supposed to retrieve and try on a new backpack—the AMU, or Astronaut Maneuvering Unit—that contained an independent supply of oxygen and some propellant for a jet maneuvering gun. He discovered that in zero gravity even trying to strap himself into the AMU required enormous effort because the slightest move sent him tumbling out of control. The 24-foot-long tether—he called it the snake—got in his way. His faceplate fogged with perspiration. Cernan became so exhausted that the space walk had to be cut short.

With the next two flights, both three-day missions, most of the troubles and frustrations that had assailed Project Gemini early in 1966 disappeared. *Gemini 10,* launched on July 18, took John Young and Michael Collins to the first trouble-free docking with the Agena.

Gemini 11 was even more successful. To simulate rendezvous in lunar orbit, Pete Conrad and Richard Gordon had to achieve a first-orbit rendezvous with the Agena. To do so they had a two-second "launch window"—they had to blast off within two seconds of schedule. On September 12, they did—with 1.5 seconds to spare. They docked with the Agena and then used its engines to power the joined vehicles to a record altitude of 850 miles. Later, with a 100-foot Dacron tether taut between them, the spacecraft and the Agena revolved slowly, creating a weak field of artificial gravity.

But the successes on these two flights were overshadowed by the problems of space-walking. Though application of a chemical to the faceplate prevented the fogging problem, Collins on *Gemini 10* got excessively tired and tangled up in the umbilical. Gordon on *Gemini 11* went outside to tether the Agena to the spacecraft and, after a few minutes, reported, "I'm pooped."

Proving that astronauts could work outside the spacecraft now became the urgent goal of the final flight, *Gemini 12,* manned by Jim Lovell and Edwin (Buzz) Aldrin. When Aldrin began his space walk on November 13, two days after launch, every provision had been made.

Aldrin's EVA was programed practically to the second and included a dozen two-minute rest periods. Aldrin was equipped with a shorter umbilical, a portable handrail and two handholds—with Velcro backing that would adhere to Velcro patches plastered over the outer skin of the spacecraft—and a waist harness like that used by window washers to hook into rings on the vehicle. Aldrin had even rehearsed his every move wearing a pressure suit in an underwater tank—conditions that approximated zero gravity.

Anchored securely by his new restraints and pacing himself carefully, Aldrin cleaned Jim Lovell's window with a cloth, then moved back to the adapter section. There, at a "busy box" of work tasks, he torqued bolts with a wrench and performed other typical space repair jobs without undue exertion. Later, he tethered the Agena to the Gemini docking bar, and returned to the cabin—barely winded after 120 minutes.

Aldrin and Lovell splashed down at 2:21 p.m. on November 15, 1966. Like the previous four flights, they were within three miles of target. Project Gemini was now ended. At a cost of $1.1 billion, this "afterthought" program had established beyond doubt the human capacity to master space. In the time that the Russians remained earthbound, 16 U.S. astronauts had spent a collective total of 1,940 hours aloft. They had answered the critical questions. Yes, pilots could rendezvous and dock with another vehicle, safely endure a two-week mission, control their return to earth and—as Aldrin proved so dramatically—even step out of the spacecraft and act as celestial mechanics. Now it was Go for Apollo. □

THE FIRST COSMIC WALK

Cosmonauts Alexei Leonov *(left)* and Pavel Belyayev wave from a flower-decked auto after the flight of *Voskhod (Sunrise) II.* The 30-year-old Leonov traveled 3,000 miles during his 10 leisurely minutes outside the capsule.

Leonov sketches Belyayev during a postflight vacation on a Black Sea beach. It was Belyayev who first radioed back to earth that "a man has stepped out into cosmic space"—a message he repeated twice in his excitement.

On March 18, 1965, cosmonaut Alexei Leonov, junior pilot aboard the *Voskhod II,* climbed into the inflatable air lock, "gave a little push and popped out of the hatch like a cork." By now the form was old hat: another Soviet spacecraft, another space first for the Soviets. But Leonov's act itself was dramatically, stupendously new. A man had actually stepped out into space, touched the heavens and been, at least for 10 minutes, a planet unto himself.

Only a 16-foot tether kept that human planet from entering what one journalist called "a fatal lonely orbit of his own." Yet the weightless Leonov insouciantly performed headstands, somersaults and minor mechanical tasks as he moved around and even in front of the capsule— once tugging himself back so hard that he bounced off again like a spring. "I felt absolutely free," he said later, "soaring like a bird as though I was flying by my own efforts."

The space walk took place early in the second orbit, while the craft was over the U.S.S.R. Leonov talked with command pilot Pavel Belyayev and with leaders and friends in Moscow (including first cosmonaut Yuri Gagarin), and he took motion pictures that were relayed back for broadcast around the world. But only Leonov experienced the pure airless void itself—the overwhelming cold of space and the powerful rays of the unfiltered sun "welded into the black velvet of the sky." Fear was unthinkable. "The view of the cosmic expanse so enthralled me that there was no place in my mind for any other feelings," Leonov reported. "I had only time to look, to be stupefied—and to carry out the program."

With earth 110 miles beneath him, Leonov floats away from the capsule, connected by the tether holding radio and telemeter lines. The white, four-layered spacesuit reflected the intense, unfiltered sunlight.

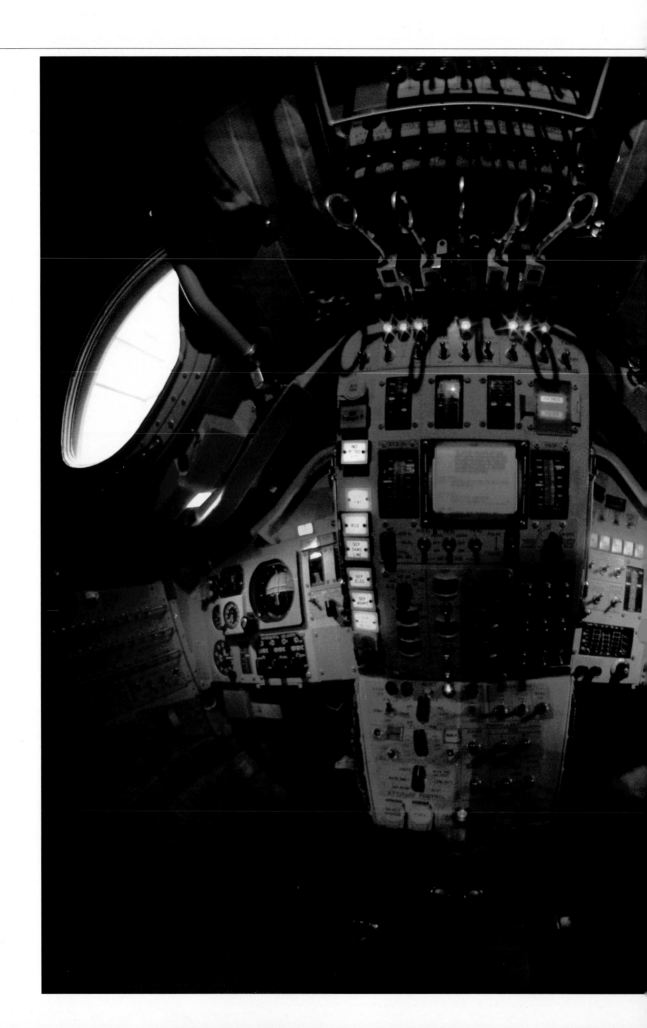

SPACECRAFT ANATOMY

Despite its cramped appearance, the Gemini spacecraft was a good 50 per cent roomier than the Mercury capsule upon which it was based. But more important than the increased space itself was the smarter (and generally simpler) use that was made of it.

Gemini could hardly have matched the awkward complexity of the Mercury capsule, which had been rushed through production with one goal in mind: to get a man into space fast. Most Mercury systems were packed into the pilot's cabin—"stacked like a layer cake," said one NASA engineer—with parts scattered about in a tangle of interconnecting wires and tubes. Fixing one bad system inevitably meant disturbing many healthy ones—and all of them then had to be thoroughly rechecked when the work was done.

By contrast, most of the Gemini systems were self-contained modular packages that could be replaced whole when they broke down. They were stowed behind hatches in the exterior wall, where they could be serviced "without hauling the crew out and turning the cabin upside down," as astronaut Gus Grissom described it. This arrangement speeded assembly as well, enabling many men to work simultaneously on the outside, rather than having one man at a time working within.

Modularization also helped eliminate what one designer called "the root of all evil" in the Mercury capsule: the automatic sequencing system that tied one function to another in a daisy chain of electronic interdependence. A failure in any one of 220 electrical relays could bring a test or a countdown to a lengthy halt. With Gemini's discrete systems, sequencing became mental, part of the pilot's—not the computer's—job. The number of failure-prone relays was thus cut by 75 per cent.

For all that, the Gemini spacecraft was still an awesomely complex piece of work—even more complex, in some ways, than Mercury. It could rendezvous and dock with another orbiting spacecraft. And where the Mercury capsule's hatch was secured with explosive bolts—designed simply to release the pilot after the mission—Gemini had a pair of hatches that could be opened and closed by the astronauts themselves when going on space walks.

The spacecraft's cockpit held a boggling array of instrument panels, indicator lights and switches. Yet there was clear order in the design. The command pilot sat on the left, working controls for guidance, rendezvous and landing. The second pilot managed the craft's computer and monitored its fuel and radar systems. The attitude-control stick sat midway between the two astronauts, each of whom had a large round dial showing the ship's exact attitude with respect to the earth. A big screen at top center projected the flight plan; gauges to its left and right measured oxygen and fuel levels. The panels "giving us information on our safety are directly before us," Grissom noted approvingly. "Those less critical are farther from our straight-ahead vision. All in all a very businesslike, nononsense layout, and the only thing they didn't include is Muzak." □

The attitude-control stick fires rocket thrusters that pitch the craft up or down, yaw it left or right, roll it upside down—or propel it into a different orbit altogether.

Rounded hatch windows frame a maze of instruments inside the Gemini spacecraft. The attitude-control stick is at bottom center. The silver rings at top regulate cabin ventilation and oxygen supply during landing.

Assembling wires for a relay panel inside the cockpit, an electrician uses liquid silicon to waterproof and insulate each wire where it enters the plug. A simplified sequencing system—and especially the elimination of the escape tower—reduced the number of relays from 220 in Mercury to 60 in Gemini.

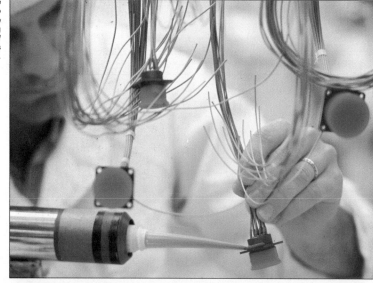

Two technicians sit inside the open hatches to install the water-supply system, while three others work on the reentry control system, whose thruster engines jut out at top. Between the two groups an infrared lamp cures the waterproofing around the radio-antenna plug.

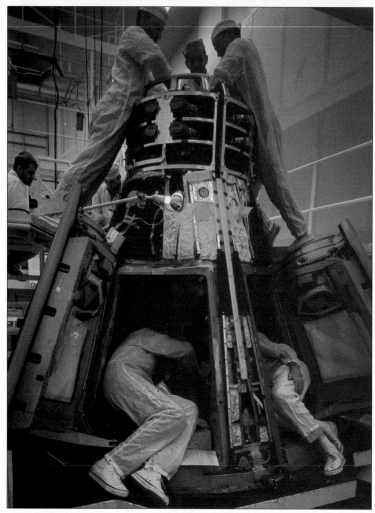

Wrapped in a protective plastic cover, the top section of the spacecraft's cabin is ready for installation. The carefully twisted wire bundles connect the guidance, computer and instrumentation systems; the numbered squares mark connections in electrical circuits.

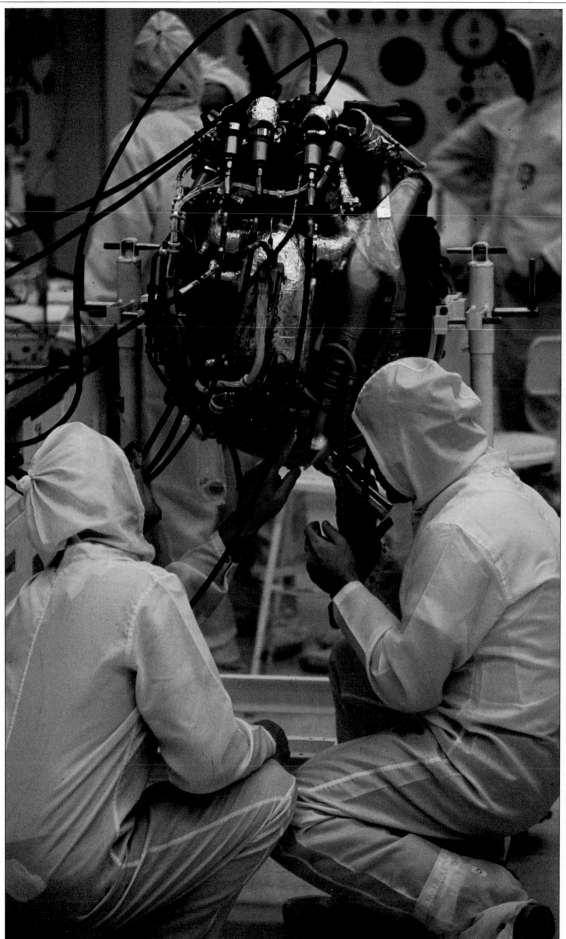

White-clad technicians test the spacecraft's environmental system, which supplied the pilots with oxygen and water. The engineers wore special boots, hoods and coveralls to keep dust or even a single hair from getting into parts.

Black arms jut from a sealed bubble used when welding pieces of titanium. The welder stood outside, stuck his hands into the inflated gloves and reached inside, where the air had been replaced with nonflammable argon gas to help avoid imperfect welds.

KRAFT IN CONTROL

Flight director Chris Kraft calls up some data during the four-day voyage of *Gemini 4*. The Houston control room, which Kraft helped design, housed four rows of 20 consoles each, facing a giant map of the spacecraft's progress. On the floor above was a second control room—so Kraft could conduct a simulated flight even as the real one went on.

Smoking one of his three daily cigars, Kraft stares up at the control room's orbit map. He was only 37 years old when he directed his—and NASA'S—first manned space flight in May of 1961. "There were no schools we could go to to learn this business," he said. "We just wrote our own textbook as we went along."

In Project Gemini, as in Mercury before and Apollo after, the "control" in Mission Control was an aeronautical engineer by the reassuring—and slightly improbable—name of Christopher Columbus Kraft Jr. As NASA flight director, Chris Kraft ran a worldwide network of tracking stations, relay satellites, rescue planes and recovery ships—as well as the control room itself. He decided when to launch or abort, monitored every aspect of the flight and ultimately took the blame if things went wrong.

Kraft thrived on the responsibility and excitement of his job. "Just before launch, my pulse rate is higher than that of the astronauts," he admitted after one of the missions. "None of us can relax until those guys are on the carrier." He seemed always to be in the control room, even when he was supposedly being spelled by one of two deputies. And he demanded like dedication from his Houston staff, which numbered nearly 600 people early in the Gemini program.

To help prepare his staff, Kraft put together the Mission Rules book—a constantly expanding compendium of procedures for handling emergencies. He also ran dozens of simulations before every launch and conducted critiques after every mission, real or practice.

Kraft built a remarkable team spirit, but could come down hard when he had to. The mission came first, and no one—including the astronauts—had any doubts about who ran the mission. "He's a virtual dictator," said one assistant, "which is the way it has to be." Kraft agreed. "In an emergency," he said, "there's no time to argue." Kraft's own description of his role likened it to that of a symphony conductor. "The conductor can't play all the instruments," he said. "He may not even be able to play any one of them. But he knows when the first violin should be playing, and he knows when the trumpets should be loud or soft and when the drummer should be drumming. He mixes all this up and out comes music. That's what we do here."

Sighting across his domain, Kraft holds the control pack, which enabled him to talk to engineers and technicians at consoles around the control room. An unflappable genius at making complex decisions—often without having much information to go on—Chris Kraft was described by a colleague as having a remarkable "ability to think in real time."

Tension showing in his face, Chris Kraft studies
his instrument-packed control room during
Gemini 4. At launch on that mission, Kraft's
pulse rose to 135—nearly double its normal rate.

NASA engineers monitor their consoles during
a Gemini rehearsal. No one—including Chris
Kraft—knew what ''emergencies'' the
computerized simulation system would create.

GEMINI

MISSION PORTFOLIO

After lift-off on March 23, 1965, Gus Grissom, commander of *Gemini 3,* and pilot John Young executed maneuvers that had never been attempted before. On the first of three orbits, Grissom used the spacecraft's 16 small thrusters to drop it from its elliptical launch orbit into an almost-circular loop—a vertical shift he would repeat just before reentry. Then, as they entered their second orbit, he shifted the craft's orbital plane by ⅟₅₀ of a degree. The ability to change orbits was a major development in space travel. For the first time the astronauts truly had control of their craft—a control that would be essential to any planned lunar landing.

Splashdown was 50 miles short of the recovery ship and the astronauts spent a queasy half hour bobbing in the closed spacecraft before the pickup helicopters arrived. Grissom—who had given *Gemini 3* the nickname *Molly Brown* as a charm against its sinking like the *Liberty Bell 7* in 1961—refused to open any hatches until the divers had attached the flotation collar. "It was a smooth, successful flight," Grissom said later. "But if *Molly Brown* had sunk out there, I'd have jumped right off that carrier."

Seen through a window of their spacecraft, Grissom *(right)* and Young wait out the countdown at Cape Kennedy. *Gemini 3's* lift-off was so smooth that neither of the astronauts felt anything—the starting of the mission clock on the instrument panel alerted them.

Gemini 3 lifts off from Pad No. 19 in an almost flawless launch. The 110-minute final countdown was interrupted only once—in order to correct a minor malfunction in the main engines.

At home outside Houston, Texas, John Young's family watches *Gemini 3's* lift-off on television. "Fantastic," reacts Barbara Young, but six-year-old Johnny, ill with chicken pox, and eight-year-old Sandy look apprehensive.

Frogmen who attached a flotation collar to *Gemini 3* watch as John Young ascends in a helicopter rescue harness. Both Young and Grissom, suffering from heat and seasickness, had doffed their spacesuits and were clad only in long underwear when help arrived.

About an hour after splashdown, Young *(left)* and Grissom—outfitted with Navy robes—are delivered by helicopter to a red-carpet reception from the crew of the carrier *Intrepid*.

Still looking a bit peaked, Young and Grissom stride across the carrier deck. The astronauts spent two more days aboard the carrier, undergoing debriefing and medical exams.

The primary objective of *Gemini 4,* crucial though it was for the future of the space program, was almost prosaic: James McDivitt and Edward White drew the basic assignment of simply waiting out the effects of a four-day trip, the longest so far by Americans. It was the more dramatic secondary goal—EVA, or extravehicular activity—that captured the imagination of the world. On June 3, 1965, during *Gemini 4's* third revolution, White became the first American to walk in space, connected to the spacecraft by a 24-foot tether providing oxygen and two-way communication with McDivitt. "I can sit out here," he noted midway through the 20-minute EVA, "and see the whole California coast." Then White rejoined McDivitt for 59 more revolutions without harm—a convincing argument for the feasibility of even longer trips.

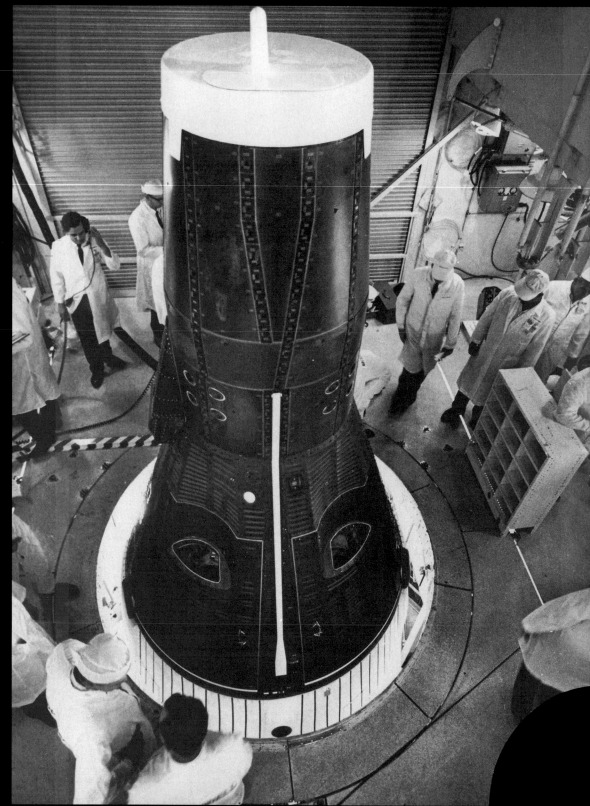

On Pad 19 at Cape Kennedy, technicians from NASA and from the McDonnell Aircraft Corporation, builders of the spacecraft, stand by for the start of countdown after sealing the crew of *Gemini 4* into their spacecraft.

After the hatch has been opened, White stands up in airless space. In his left hand he holds a "zip gun," which he fired in bursts to help him maneuver outside the spacecraft. Mounted atop the maneuvering unit is a camera.

The space walker's tether uncoils as he pushes away from the spacecraft with a burst from his gun—which ran out of its compressed-oxygen fuel in a few minutes. Thereafter, White tugged on the tether to move around.

of the spacecraft. On the hatch window
(lower right) is the shadow of the movie camera
that took these pictures. In case oxygen
flow from the umbilical was cut off, White had an
emergency supply in a pack on his chest.

White pulls himself around the outside of the spacecraft *(left)* to give McDivitt a better angle for this through-the-window photograph. Then he parked himself outside the window *(below)* for the duration of the EVA.

RE OF

...ng out,
...some talking to get
...in.

...want you to get

...*g):* I'm not coming
...un.
...e on.
...come back to you but

...n, you still got three
...re days to go,

...ROL: You've got
...utes to Bermuda.
...g to . . .
...y, come in then.
...n't you going to
...?
...come on in the . . .
...n here.
.... I'll open the door
...ough there . . .
...e on. Let's get
...before it gets dark.
...saddest moment of

...you're going to
...when we have to
...ith this thing.
...ROL: *Gemini 4* . . .
...g to come in the

...ROL: *Gemini 4.* Get
...ou getting him back

...standing in the seat
...egs are below the
...nel.
...ROL: Okay, get him
...e going to have

White gives the thumbs-up sign as he and McDivitt arrive on the deck of the carrier *Wasp*. Neither showed any ill effects from their 98 hours aloft, but both had lost weight; White shed about eight pounds, McDivitt about four.

"Eight days or bust" was the covered-wagon-era slogan that Gordon Cooper and Pete Conrad chose for *Gemini 5*, which lifted into orbit on August 21, 1965. They made it all the way, despite a plague of troubles that sometimes threatened to abort—or gravely hamper—the mission. On the first day, balky fuel cells, the spacecraft's main source of electricity, reduced power to worrisome levels. And on the fifth day, thrusters malfunctioned; unable to keep the spacecraft on an even keel, the astronauts spent three days drifting and slowly tumbling through space. Both of these deficiencies corrected themselves in time for completion of the flight—but then a computer error dumped the craft in the Pacific 103 miles from its recovery ship. Nonetheless, *Gemini 5's* 190-hour mission broke all the records and proved the astronauts equal to the rigors of extended space travel.

Conrad waits through countdown inside the spacecraft. *Gemini 5's* troubles began before liftoff. Launch was delayed 48 hours because of thunderstorms and a lightning strike near the launch-pad power supply; technicians were afraid that the strike might have caused damage to the spacecraft's computer.

Signal dye colors the water around *Gemini 5* as a frogman wearing scuba gear jumps from a helicopter to help with the craft's recovery after splashdown on the 29th of August, 1965. In the distance, the destroyer *Du Pont* stands by.

Conrad is hoisted in a sling to the helicopter that flew him and Cooper to the *Lake Champlain*, 75 miles away. The frogmen stayed with the spacecraft to await a later pickup by the carrier.

Four hours after splashdown, the carrier *Lake Champlain* eases toward the spacecraft, which bobs gently on its flotation collar. Though *Gemini 5* landed on a calm day, rescue teams as well as astronauts were trained to cope with heavy seas—even sharks—if necessary.

Aboard ship, Conrad mugs as he tugs at Cooper's eight-day beard. Watching the scene on TV in Houston, Conrad's wife, Jane, quipped: "Frankenstein meets the Wolf Man."

All the orbital jockeying of earlier missions had been leading to a major goal: the rendezvous of two vehicles in space. At 8:37 a.m. on December 15, 1965, Walter Schirra and Thomas Stafford in *Gemini 6* rose from the launch pad in pursuit of Frank Borman and James Lovell in *Gemini 7,* launched 11 days earlier. Once in orbit, Schirra used delicate bursts of his thrusters, first to change the elliptical launch orbit to a circular loop, then gradually to close on his quarry as he made the two orbits identical. At 2:33 p.m., *Gemini 6* had closed to within 120 feet of her sister ship, with no relative motion between them. On the ground, cheering flight controllers and Mission Control chief Chris Kraft waved miniature American flags and lit up triumphant cigars: The world's first manned rendezvous in space had been achieved. For almost five hours, the two ships flew together, sometimes no more than a foot apart. Then, after all four astronauts had had a nap, *Gemini 6* headed home and *Gemini 7* went on to complete its own mission by remaining in orbit for a total of 14 days, a new endurance record.

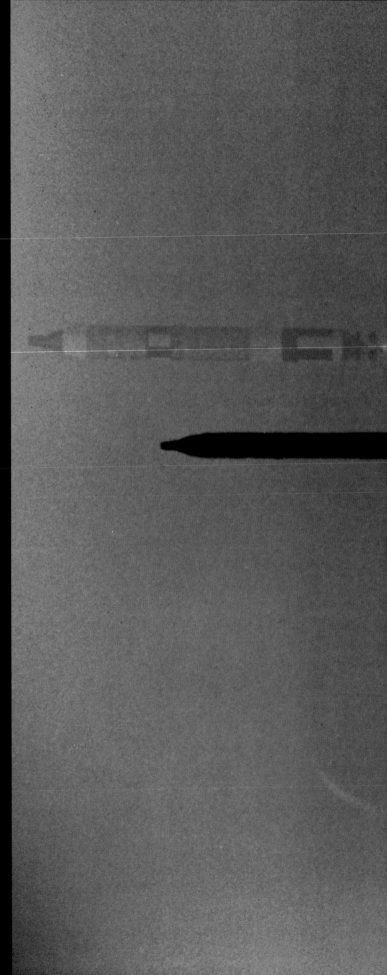

In a double exposure made from different vantage points north and south of the launch pad at Cape Kennedy, *Gemini 7*'s ghostly shape seems to lead the way for *Gemini 6*, whose original lift-off had been scrubbed—and then rescheduled to let *Gemini 7* go first.

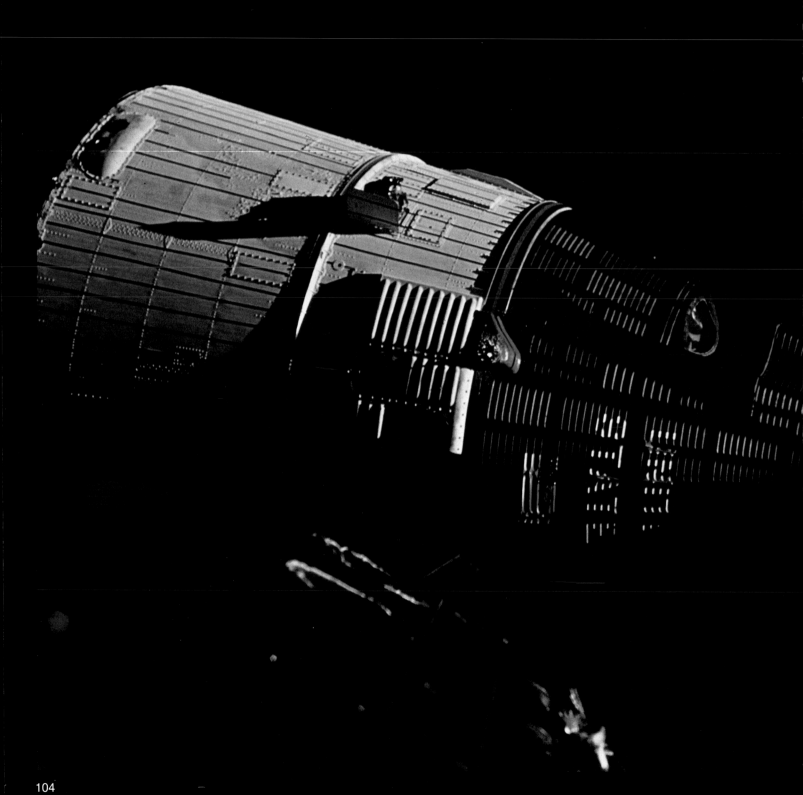

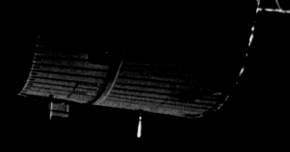

Viewed from *Gemini 6*, dawn in space
touches the rear section of *Gemini 7 (above)* with
a faint glow, a scene repeated every 96
minutes throughout the rendezvous. The straps
at the back are tapes ripped loose at launch.

At left, *Gemini 6 (foreground)* nudges so
close to *Gemini 7* that Lovell's face can be seen
through the right-hand window. "Our nose
continually pointed at them, as we moved around
their midriff," *Gemini 6* pilot Schirra
reported later. "The thrill was fantastic."

Sunlight glares off the white equipment
section of *Gemini 7*, where most of the flight gear
is housed. Visible in the nose of the
spacecraft are two circular radar receivers and a
larger, round parachute installation.

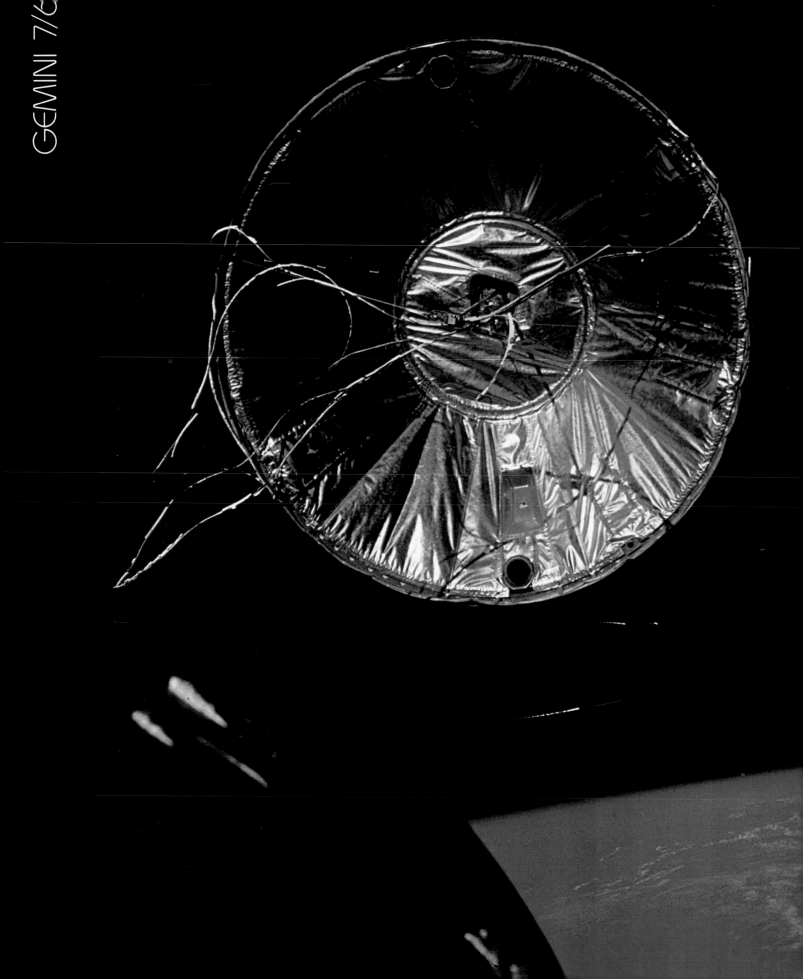

Inside *Gemini 7*, Borman performs a test designed to determine the effects of prolonged space travel on astronauts' vision.

In a photograph by Borman, Lovell's face shows the strain of the long flight—and perhaps the added stress of imminent parenthood: He and his wife, Marilyn, were expecting a baby.

After splashdown, Stafford emerges from
Gemini 6's left hatch and Schirra from the right
to savor fresh air again while a frogman *(far
right)* uses a radio phone to summon the pickup
helicopter and the recovery ship.

The flight of *Gemini 8* had been almost flawless: Neil Armstrong and David Scott, aiming for the first docking in space, had lifted off right on schedule on March 16, 1966. The rendezvous with the unmanned Agena target vehicle was routine, the docking maneuver a triumph. Then the venture unraveled. Joined together, the two craft began simultaneously rolling and tumbling end over end at a rapidly increasing rate. Armstrong tried to stabilize them by using his maneuvering thrusters, failed—and undocked. But once free of *Agena*, *Gemini 8* gyrated faster. "We have serious problems here," Scott radioed Mission Control. In a final effort, Armstrong deactivated the maneuvering system and fired the reentry maneuvering rockets. The spinning stopped. A quick test showed that the No. 8 maneuvering thruster had been stuck in the "on" position. But the premature firing of the reentry rockets required an immediate splashdown to ensure that the craft would have enough fuel for reentry.

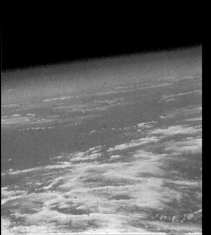

In a sequence taken by Scott during *Gemini 8's* dusk approach, *Agena* looms ever larger until only a few feet separate it from the nose of the spacecraft. *Agena 8's* red and green running lights were turned on from the ground.

Highlighted by the sun, the unmanned Agena target vehicle seems to hang motionless above the clouds of earth. An eight-foot radar antenna rises just aft of the docking cone, which is fitted out to receive *Gemini 8's* nose.

Scott's camera catches *Agena* at the moment Armstrong separated the two vehicles. Neither *Gemini's* maneuvering thrusters nor the remote control of *Agena's* systems could stop the menacing spin, so Armstrong undocked.

Gemini 8 spins dizzily at one revolution per second. The emergency occurred while the craft was out of ground-communications range; otherwise, ground-controlled telemetry could have diagnosed the problem at once.

Gemini 8's wild spin in the sunlight registers on film as a red and yellow blur. The astronauts were already struggling to see their instruments when they decided to use the re- entry thrusters to steady the spacecraft.

After their unscheduled descent, Armstrong *(above, left)* and Scott calmly await pickup, with the hatches open despite a choppy sea. *Gemini 8* came down at an emergency landing point 621 miles south of Yokosuka, Japan.

Armstrong and Scott chat with a well-wisher back at Cape Kennedy. Their spacecraft was returned to the McDonnell factory, where engineers discovered it was a short circuit that had caused the thruster to malfunction.

STAFFORD · CERNAN

Beginning with eleventh-hour postponements that stalled *Gemini 9* twice before it got off the ground on June 3, 1966, the mission was plagued by mishaps. The priorities for Thomas Stafford and Eugene Cernan were rendezvous and docking, followed by an EVA with a Buck Rogers-style backpack designed to let Cernan maneuver independent of the spacecraft. But docking proved impossible: A shroud to protect the target vehicle's docking cone failed to detach after launch—creating jaws that looked, in Stafford's words, "like an angry alligator." Cernan's EVA was equally frustrating. The effort to don the backpack exhausted him and overloaded his spacesuit's cooling ability. Just as he managed to put the maneuvering unit on, mission controllers recalled him. Perspiration had fogged his faceplate, and Cernan could not see. Although he was outside a record two hours and eight minutes, that was 39 minutes less than planned. But *Gemini 9* redeemed itself on splashdown, landing so close to the recovery ship that, according to one newspaper, any closer and the carrier would have had to duck.

Gemini 9's inaccessible target orbits with its launch nose shroud still attached by wires that had been mistakenly taped down. Unable to dock, the astronauts improvised—using the vehicle to practice rendezvous maneuvers

With a triumphant thumbs-up for cheering sailors on the *Wasp*, the astronauts wait for the carrier to pick them up. *Gemini 9*'s splashdown was so close to target that the frogmen found the craft still hot to the touch.

A successful docking with an Agena target vehicle was the key to a series of unique space achievements for John Young and Michael Collins in *Gemini 10.* After launch on July 18, 1966, the astronauts overtook and docked with *Agena 10* on the fourth orbit. Later they used *Agena 10's* 16,000-pound engine to boost themselves to a record orbital apogee of 474 miles. After dropping back into a lower orbit, *Gemini 10* set off for an unprecedented second rendezvous. The target was the derelict *Agena 8,* still orbiting but unable to emit radar signals; Young and Collins would have to make the delicate final closing maneuvers using only optical equipment. En route, Collins stood up in the open hatch to take photographs for a study of stellar ultraviolet radiation. Then, after a sleep period, Young undocked from *Agena 10* and made the rendezvous with *Agena 8*—the first accomplished without the use of on-board radar. Collins then retrieved an experiment package from *Agena 8,* becoming the first astronaut to perform EVA twice on one mission.

Agena 10 glints in a harsh sunrise in this photograph taken by Collins from 100 feet away. The astronauts were launched only 100 minutes after *Agena 10,* when a 35-second "launch window" offered them the best chance to achieve rendezvous with both targets.

Gemini 10 approaches *Agena 10* in final docking maneuvers *(below)* in the fifth hour after launch. With the vehicles firmly coupled at five hours and 52 minutes into the flight, the view from Collins' window *(right)* shows an instrument panel on *Agena 10's* nose, just below the radar antenna. The panel provided such data as the target vehicle's rocket-fuel levels.

Still in their spacesuits after splashdown, Young *(left)* and Collins show weariness from their demanding, complex mission. Though the astronauts did manage to take a brief nap nine hours into the flight, their first real sleep did not come until 30 hours had gone by.

"We're over Australia now," command pilot Pete Conrad radioed from *Gemini 11.* "We have the whole southern part of the world at one window. Utterly fantastic." Conrad and Richard Gordon had docked with *Agena 11* just 85 minutes after their lift-off on September 12, 1966, and had immediately set out to improve on the high-riding record of *Gemini-Agena 10.* They beat it with ease—by a whopping 376 miles.

Although Gordon had to cut short his EVA when he was blinded by sweat in his right eye, he helped to accomplish one highly important objective by attaching a 100-foot-long Dacron tether to link *Gemini 11* and the Agena target. Conrad then undocked from *Agena 11* and backed slowly away, stretching the tether taut. At first, the vehicles wobbled and the tether swung like a skip rope. Firing a burst from his thrusters, Conrad stabilized the gently rotating spaceships, generating a mild centrifugal force that simulated a gravitational pull. Loose objects in the astronauts' cabin began drifting toward the rear wall— where they finally "fell" in the newly created artificial gravity.

A photograph taken from a record 850 miles up covers the northwest coast of Australia. To the tracking station at Carnavon, Australia, Conrad later exclaimed, "I've got India in the left window, Borneo under our noses, you're in the right window—and the world is round!"

Sporting mission caps presented after their recovery, Gordon *(left)* and Conrad maintain a mood of exultation over the success of their flight in spite of Gordon's EVA troubles.

Aboard the *Wasp* after splashdown, Aldrin *(left)* and Lovell grin in the triumph of a successful finale. This flight, in combination with his 14 days aloft in *Gemini 7*, gave Lovell a total of more than seven million miles in orbit.

APOLLO TO THE MOON

Even as Project Gemini scored success after success during 1966, preparations for Project Apollo were being pushed to meet the national commitment of a lunar landing before the end of the decade. While the Soviets had launched no manned flights during the 20 months of Gemini missions, they had recently sent two unmanned probes to the moon and its vicinity. This show of interest, U.S. space experts suspected, might be the prelude to a Soviet attempt to put cosmonauts on the moon before Apollo could get off the ground.

Rushing ahead, NASA scheduled the first Apollo manned flight for February 1967, only three months after the final Gemini mission. The three-man crew for *Apollo 1* consisted of veterans Gus Grissom and Ed White and rookie Roger Chaffee who, at the age of 31, was the youngest astronaut selected to go into space.

On January 27, about three weeks before the flight—which was to test the new spacecraft in earth orbit—the three astronauts were seated high above Pad 34 at the Cape, sealed in the cockpit of *Apollo 1* for a practice countdown and simulated launch. This rehearsal was a "plugs-out" test—in other words, the umbilical connections from the ground were removed from the spacecraft and its Saturn booster as they would be during lift-off. The test had been interrupted repeatedly that afternoon by a succession of glitches in the communications system. In fact, Grissom had been grumbling lately about problems with the spacecraft and had hung a lemon on it—a Texas-sized lemon picked from the tree in his yard in Houston.

At 6:31 p.m. one of the astronauts, probably Chaffee, announced almost casually over the intercom, "Fire. I smell fire." Then the calls became more insistent. A television camera trained on the small window of the spacecraft hatch showed indistinct motion of legs and feet and a sudden flash of flame. A sharp cry came from the spacecraft, then silence. About 14 seconds after the first alarm, the overheated capsule ruptured, belching fire and black smoke. Technicians rushed to pry open the hatch, but it was nearly six minutes before they succeeded. Grissom, White and Chaffee were already dead of carbon monoxide asphyxiation in what should have been nothing more than a routine test.

The disaster on Pad 34 stunned a nation accustomed to hearing only good news from its space program. While the very future of Apollo hung in the balance, hundreds of government investigators sifted through the charred remains of the spacecraft. They were unable to pinpoint the exact cause of the tragedy but concluded that the fire had started in or near a bundle of wires in front of Grissom's couch on the left. One of the wires apparently had sparked, igniting flammable materials that burned explosively in *Apollo 1's* highly pressurized pure-oxygen atmosphere.

In their 3,000-page report on the accident, the investigators were highly critical of both NASA and the spacecraft's builder, North American Aviation (which became North American Rockwell in September 1967), citing hazardous test procedures, inadequate safety precautions and deficiencies in the spacecraft design and workmanship. The report led directly to a number of changes in the Apollo spacecraft. Wiring was rerouted and shielded with better insulation. Combustibles, including the nylon covering of the astronauts' spacesuits, were replaced with flameproof materials.

The cockpit's atmosphere, pressurized on the ground to keep out the Cape's salt air, was changed to a less volatile two-gas mixture—60 per cent oxygen, 40 per cent nitrogen. After lift-off, pressure would be lowered and the atmosphere gradually converted to pure oxygen. Finally, to make escape easier, engineers devised a hatch that opened outward and could be unlatched by the astronauts in three seconds; *Apollo 1's* hatch swung inward and required at least 90 seconds to open.

Next, NASA threw out its timetable for missions and began a complete overhaul of the entire Apollo system—testing, correcting and retesting each component in a process that took many months. Apollo had three major parts, and each one needed to be checked individually: the command and service module (CSM), consisting of the 10-foot-high cockpit and the adapter section (which included the propulsion system for maneuvering en route to the moon); the lunar module, or LM *(pages 130-131);* and the three-stage Saturn V booster, with its 7.5 million pounds of thrust at lift-off.

The booster alone was so tall—364 feet—that it had to be assembled on a mobile launch platform in a specially constructed building. Then it had to be trucked—standing straight up—to either of a pair of launching pads on a tracked crawler-transporter that was the largest land vehicle ever built. This mobile launch concept made it possible for NASA officials to alternate the use of the two pads and to schedule lift-offs as often as every two months.

The Apollo flight schedule called for seven different types of mission—progressively more difficult and assigned the letters *A* through *G*—to test the reliability of all the systems. Letter *G* stood for the climax, the lunar landing, but the schedule began with *A* and *B,* a series of five unmanned flights to check out the various components. The most important of these unmanned flights occurred on November 9, 1967: *Apollo 4* saw the first launch of the CSM by the big Saturn V booster.

Meanwhile, the corps of astronauts who hoped to fly to the moon had expanded again. The first six scientists were added and then a group of 19 pilots, bringing the grand total of men selected for space flights to 55 thus

far. There also had been losses. Two of the original Mercury men, John Glenn and Scott Carpenter, had retired from the program and, in addition to the three victims of the *Apollo 1* fire, five astronauts had died in other kinds of accidents not associated with space flight.

As crews were assigned to each mission, their training concentrated on endless sessions in highly sophisticated simulators of the command and lunar modules. Some of the Apollo crews logged as many as 2,000 hours in these devices. And formerly routine geological field trips took on a particular urgency as the time drew near when some of these men might actually have to sort rocks on the lunar surface.

It was October 11, 1968—nearly 21 months after the fire—before officials felt ready to launch the first manned flight, *Apollo 7*. On board were a pair of rookies, Walter Cunningham and Donn Eisele, and oldtimer Wally Schirra who, at 45, had announced his impending retirement before the flight. Because the mission called for testing only the CSM in earth orbit, they rode aloft on a smaller booster, the two-stage Saturn 1B. In orbit, they demonstrated Apollo's ability to rendezvous with the second stage of the booster, then settled down for nearly 11 days in space— with Schirra hosting seven live television transmissions from "the lovely Apollo room high atop everything."

So nearly perfect that ground controllers called it "101 per cent successful," this maiden flight set the stage for a remarkable demonstration of the Apollo program's flexibility. *Apollo 8* had been scheduled to test the lunar module in earth orbit, but the LM was not yet ready to fly. So the schedule was juggled and, on the 21st of December, 1968, *Apollo 8's* crew of Frank Borman, Jim Lovell and William Anders became the first astronauts to ride the gigantic Saturn V booster into space. Their mission: to break the grip of earth gravity and

Apollo 8, the first manned spacecraft to orbit the moon, lifts off at dawn on December 21, 1968—its destination artfully placed above the fire storm in this composite photograph. In reality, however, the new moon had already slipped below the horizon when the towering Saturn V rocket boosted *Apollo 8* skyward.

to fly around the moon and back.

Their 240,000-mile journey took nearly three days. On Christmas Eve, as *Apollo 8* dipped to within 60 miles of the lunar surface, sending back live TV pictures of that stark and alien place, there came a message:

In the beginning God created the Heaven and the Earth.

Nearly a quarter of a million miles from home, Bill Anders was reading from the Book of Genesis. Then Lovell took up the reading:

And God called the light Day, and the darkness He called Night.

Borman, the mission commander, completed the familiar verses:

And God called the dry land Earth; and the gathering together of the waters He called seas; and God saw that it was good.

On Christmas Day, Borman fired the service module's engine to break free of lunar gravity for the return trip home, a maneuver both precise and dangerous. But the operation was almost anticlimactic for the estimated half billion people who had listened to the broadcast the day before. It had been a year of trouble, of war in Vietnam and dissent in the United States, and those few minutes of rare emotional drama had helped redeem it. "You have bailed out 1968," a friend telegraphed Borman.

Now, almost two years after the fire had thrown the entire project into question, Apollo was back on track. However, one crucial component, the lunar module, still had to be tested in flight. This was the assignment of *Apollo 9* and its crew of Jim McDivitt, David Scott and Russell Schweickart, who went into earth orbit on March 3, 1969.

Their 10-day mission was the riskiest thus far. First, they separated their command and service module from the third stage of the Saturn rocket and pivoted in space. Next, they docked with the lunar module *Spider,* plucking it from the nose of the third stage, where it had been stowed with legs folded during the launch *(page 131).* Then, after Schweickart had crawled into the LM and made a successful space walk from it, he and McDivitt both entered *Spider,* fired up its engines and flew free of the command module *Gumdrop* for the first time. They maneuvered the lunar module into a different orbit and fell as much as 100 miles behind Scott, who remained alone in the mother ship.

After six hours of separation, *Spider* rendezvoused with *Gumdrop.* Delighted to see his comrades approaching, Scott radioed from the command module, "You're the biggest, friendliest, funniest-looking spider I've ever seen."

The complete Apollo system had proved itself spaceworthy and was now ready for a full-scale dress rehearsal. Launched on May 18, 1969, with a veteran crew of Tom Stafford, John Young and Gene Cernan, *Apollo 10* was scheduled to do everything except actually touch down on the lunar surface.

Orbiting the moon on the fourth day of the mission, Stafford and Cernan entered the lunar module *Snoopy* and separated it from the command module *Charlie Brown,* which was piloted by Young. While *Charlie Brown* remained in orbit 69 miles above, Stafford and Cernan fired *Snoopy's* descent engine and swooped down to within less than nine miles of the boulder-strewn moonscape—so close, as Stafford said, "all you have to do is put your tail wheel down and we're there."

En route home, the crew scored another space first: "scientific experiment Sugar Hotel Alpha Victor Echo," or shave. NASA had spent $5,000 trying unsuccessfully to develop an electric shaver that would suck up loose bristles to keep them from clogging spacecraft instruments, but this trio got satisfactory results with a drugstore safety razor and brushless lather.

Even while *Apollo 10* was speeding to the moon at nearly 25,000 miles per hour, the crawler carrying the Saturn V booster for the next launch was lumbering toward Pad 39A at a majestic .5 mph. Less than two months later, when *Apollo 11* began its epic journey on July 16, 1969, the world was waiting.

At the Cape, the contingent of VIPs and press totaled nearly 20,000—including 233 senators and members of Congress as well as media representatives from no fewer than 56 nations. When the Saturn V's five F-1 engines ignited and *Apollo 11* lifted off, these dignitaries cheered and yelled—and cried. Lift-off was so gentle that Neil Armstrong, Mike Collins and Buzz Aldrin—strapped to their couches inside the spacecraft—knew they were moving only when the altitude and velocity figures began to build up on their instrument displays. It was a Go for the moon.

". . . 75 feet, things looking good . . . lights on . . . kicking up some dust . . . 30 feet . . . drifting to right a little. . . ." The voice was Aldrin's on the fifth day of the mission. He was reporting from the lunar module *Eagle* as it descended toward the region of the moon known as the Sea of Tranquillity.

"Contact light! Okay. Engine stop," said Aldrin.

Then came the matter-of-fact voice of Armstrong, the 39-year-old mission commander, who had calmly taken over manual control of the spacecraft in order to prevent its setting down in a boulder-strewn crater: "Houston. Tranquillity Base here. The *Eagle* has landed."

It was 4:17 p.m., July 20, 1969. Seven hours later, Armstrong began the descent of *Eagle's* nine-step ladder, pausing at the second step to pull a ring that deployed a TV camera. Then, at 10:56 p.m., Neil Armstrong planted the ribbed sole of his left boot in the lunar dust.

That dramatic first step on another world turned to leaps and bounds as Armstrong was joined by Aldrin, who "felt buoyant and full of goose pimples." Cavorting in gravity only one sixth that of earth's, they spent more than two hours outside on the eerie tan-and-gray lunar surface. They set up scientific instruments, collected rock samples and planted an American flag with a telescoping arm that made the banner appear to wave in the lunar vacuum. They also placed a small memorial to the three astronauts who would never fly to the moon—one of the shoulder patches that Grissom, White and Chaffee had intended to wear on their mission.

After a nap in *Eagle's* hammocks, Aldrin and Armstrong blasted off and rejoined Mike Collins in the command module *Columbia* for the uneventful trip home—where the crew then spent 17 days in quarantine to make certain they had not brought back some unearthly microbe.

With more than five months to spare, the United States space program had fulfilled John F. Kennedy's pledge. On the grave of the slain President in Arlington National Cemetery a small bouquet of flowers appeared, along with a note: "Mr. President, the *Eagle* has landed."

The first lunar landing was a hard act to follow, but *Apollo 12* carried Pete Conrad, Dick Gordon and Alan Bean through a spectacular curtain raiser. Precisely 36 seconds after liftoff into a rainstorm on November 14, 1969, lightning struck the spacecraft, blacking out all electrical power. "We had everything in the world drop out," radioed Conrad.

Backup batteries took over until full power was restored, and Conrad and Bean went on to a near-perfect landing in the Ocean of Storms, about 1,300 miles west of the *Apollo 11* touchdown. Later, the two astronauts moon-walked nearly three times as long as their predecessors. They set up a miniature observatory that automatically transmitted data on geophysical phenomena such as lunar magnetism. In addition, they collected not only lunar rock samples but also some pieces of an unmanned Surveyor probe that had beaten them to the moon by two years.

Four men had now walked on the lunar surface. Like all feats of space exploration once replicated, voyages to the surface of the moon were beginning to seem almost routine. When *Apollo 13* was launched on April 11, 1970, the only apparent novelty was a last-minute substitution in the crew. Jim Lovell and Fred Haise were joined by Jack Swigert, a replacement for Ken Mattingly, who had been exposed to measles and lacked immunity to the disease.

En route to the moon on the third day, Lovell and Haise made a live TV transmission, showing off their lunar module *Aquarius* and letting their tape recorder float weightlessly while it played the popular song by that name from the musical *Hair*. A few minutes later, they heard a loud bang and Swigert, in the command module, saw a warning light come on. "Okay, Houston," Swigert radioed. "Hey, we've got a problem here."

A short circuit had touched off an explosion in oxygen tank No. 2, blowing out the side of the service module and disrupting the fuel cells that powered the spacecraft. With their command module crippled, the crew had to take refuge in the LM; *Aquarius* became their lifeboat.

Never had American space travelers faced such a crisis—adrift in a craft not designed for this kind of emergency, more than 200,000 miles from earth, on a trajectory that—if not altered—would leave them stranded in orbit around the moon.

With instruction from the ground, where each step was first carefully worked out on computers, the crew used the descent engine of *Aquarius* to change course to a so-called slingshot trajectory that would whip them around the moon and back to earth. Mission Control also helped them find a way to stop the dangerous build-up of carbon dioxide in the LM by suggesting a jury-rigged system of plastic bags, cardboard and tape to connect with large canisters of lithium hydroxide in the command module that could cleanse the air they breathed in *Aquarius*.

Hurtling toward earth—shivering in temperatures that dipped close to freezing, so dehydrated from conserving water that Lovell alone lost 14 pounds—the astronauts made one final course correction with the *Aquarius* descent engine. This gave them the proper angle for reentry so they would not bounce off the earth's atmosphere and go into eternal orbit.

Then they crawled back into the cold and dark command module and recharged its reserve batteries from

Apollo missions explored six different areas of the moon, as seen on the map at left. The sites of the last three expeditions were chosen primarily to increase knowledge of the moon's light-colored highland areas, created during an extremely violent phase of lunar evolution.

Aquarius' power. About an hour before reentry, they blasted loose from their lifeboat and splashed down safely in the Pacific. Afterward, when it was possible to joke about the near catastrophe, *Aquarius'* builder, Grumman Aircraft, sent a bill for towing to North American Rockwell, the command module contractor.

Investigators were quick to pinpoint the cause of the oxygen-tank explosion—defective switches had overheated during prelaunch tests and had damaged wiring insulation. But the necessary modifications delayed the launch of *Apollo 14* until January 31, 1971. To prevent a recurrence of the *Apollo 13* crew's prelaunch exposure to contagious diseases, the team of Alan Shepard, Stuart Roosa and Edgar Mitchell was kept in isolation for three weeks prior to lift-off.

Public attention during *Apollo 14's* mission was focused on Shepard, who, a decade before, had become the first American to go into space. He had been grounded by an inner-ear disorder, but that had been corrected by surgery, and at 47, Shepard was making a comeback.

To help carry geological tools and rock samples during their 10 hours of moon-walking, Shepard and Mitchell used a ricksha-like handcart. The cart allowed the astronauts to make an expedition almost to the rim of a 400-foot crater to search for the oldest rocks possible.

For *Apollo 15,* which lifted off on July 26, 1971, the handcart was replaced by the first moongoing jalopy. While Alfred Worden orbited the moon in the command module, David Scott and James Irwin cruised the surface in the Lunar Roving Vehicle *(pages 134-135),* which made it possible for them to explore up to six miles from their lunar module. Once back in lunar orbit, the astronauts released a 78.5-pound satellite to circle the moon and measure its magnetic field. And on the way home, Worden took a space walk—the first on a lunar voyage—to retrieve film

from a package of scientific experiments in the service module.

Apollo 16, which went up on April 16, 1972, more than made up for any lack of drama by the intensity of its scientific work. While Ken Mattingly piloted the command module, John Young and Charles Duke collected so many rocks—213 pounds—that they nearly ran out of bags. "By golly," said Young after splashdown, "you taxpayers—we taxpayers—got your money's worth."

The last of the missions, *Apollo 17*—carrying Gene Cernan, Ronald Evans and scientist-astronaut Harrison Schmitt—was the first to be launched at night, lighting up the entire Cape at 12:33 a.m., December 7, 1972. At 301 hours, it was the longest mission, with the longest time spent on the moon (75 hours) and the longest time spent outside the lunar module (22 hours).

Before leaving the moon, Cernan and Schmitt placed eight explosive charges scheduled to go off later. These explosions, along with the impact of the lunar module after it was jettisoned, were recorded as part of a seismographic experiment. "The Apollo program," wrote one space historian, "was leaving the moon with nine bangs and no whimpers."

If the $25.5 billion cost of Apollo was staggering, so too were its achievements: 11 manned flights, during which 24 astronauts orbited the moon and an even dozen of them left their bootprints in the lunar dust, bringing home nearly half a ton of rock and soil that proved to be up to 4.6 billion years old. But Apollo also brought home something less tangible, though no less important: A new appreciation for the beauty and fragility of spaceship Earth as it hurtles through the cosmos. To see our own blue planet through the moon voyager's eye and camera, as poet Archibald MacLeish wrote, "was to see ourselves as riders on the earth together, brothers on that bright loveliness in the eternal cold." □

THE MOON ROCKET MAN

In 1916, when space travel existed only in the fertile imaginations of such writers as Jules Verne and H. G. Wells, the Smithsonian Institution in Washington, D.C., received a remarkable document. It was titled *A Method of Reaching Extreme Altitudes*, and its pages were filled with test results, data and formulations for building a rocket to reach the upper limits of the earth's atmosphere. Only on the last page did the author, a 33-year-old physics professor named Robert H. Goddard, mention that such a rocket could conceivably go all the way to the moon.

The notion of firing a rocket to the moon captured the public fancy when the report was published in 1919. To his dismay, Goddard became known coast to coast as the "Moon Rocket Man." A songwriter dashed off a ditty called *Oh, They're Going to Shoot a Rocket to the Moon, Love!* and scores of would-be astronauts volunteered for the journey.

Despite the popular attention, Goddard's work was misunderstood and ridiculed. A *New York Times* editorial berated him for lacking "the knowledge daily ladled out in high schools." Equally discouraging was his inability to find adequate funding. In March 1920, he vented some of his frustrations—and his visions —in a letter to a Smithsonian official. The ideal rocket for space travel, he wrote, would use a highly efficient fuel mixture of liquefied oxygen and hydrogen and

could carry a human "operator" not only 240,000 miles to the moon, but 49 million miles to Mars. The rocket could be built, Goddard tartly noted, using only "well-established physical principles."

Goddard had been intrigued by rockets and the prospect of space travel long before he sent off his sensation-causing document to the Smithsonian. He was experimenting with solid-fuel rockets as far back as 1907, and in 1914 he obtained the first of 214 patents covering virtually every phase of rocket construction. But not until 1926 did Goddard succeed in launching a crude prototype of a new rocket generation. Powered by a fuel mixture of gasoline and liquid oxygen, it rose 41 feet above a cabbage patch in Auburn, Massachusetts, and crashed to earth 184 feet away.

For the next 15 years, Goddard built a variety of rockets, only to meet, ultimately, with failure. The highest that a Goddard rocket would ever fly was 4,215 feet. In Germany, meanwhile, a team of scientists was developing its own liquid-fueled rocket—the wartime V-2, which would soar 68 miles into the stratosphere. After the War—and only a few months before his death in August 1945—Goddard inspected components of a captured V-2 rocket and noted its similarities to his own work. "It's just a matter of imagination how far we go with rockets," he said. "You haven't seen anything yet."

At a classroom blackboard in 1924, Goddard shows how a properly aimed rocket with sufficient velocity—a then-unattainable seven miles per second—could escape the earth's gravity and speed through space to the moon.

In Auburn, Massachusetts, the first liquid-fuel rocket gets a prelaunch check from a well-bundled Goddard in March 1926. The rocket motor extends above the launch frame; tubing connects it to the propellant tanks below.

Like most ideas whose merit has been proved beyond doubt, lunar-orbit rendezvous (LOR)—the concept that enabled 12 astronauts to take a stroll on the moon—now seems eminently logical, if not downright obvious. The idea called for a spacecraft that would blast off from earth and go into orbit around the moon while a smaller, separate craft would detach and descend to the lunar surface; later, the small craft would lift off to return the crew to the mother ship for the trip home.

But in early 1961, when no American astronaut had yet orbited the earth, LOR was not even in the running. Most NASA officials favored either earth-orbit rendezvous (EOR) or direct ascent. Both had drawbacks *(pages 72-73)*, not least of which was that they were unlikely to meet the end-of-decade deadline for a moon landing. Direct ascent required not only the biggest booster on the drawing boards, but one that was unlikely to be ready in time. EOR required two rockets and great quantities of fuel.

Both called for landing the entire spacecraft on the moon and then relaunching it for the journey home.

Yet when a NASA engineer, Dr. John C. Houbolt, presented the concept of LOR, the reaction he got was far from cordial. "Your figures lie!" shouted Mercury designer Maxime Faget. A more subdued Wernher von Braun simply shook his head, saying, "No, that's no good."

But Houbolt, who headed a small NASA study group on space rendezvous, did not give up easily. As far as he was concerned, the advantages of LOR were clear. A moon shot would require just one rocket; and a detachable lunar lander—which would be jettisoned before return to earth—would minimize the fuel needed by the spacecraft itself.

"LOR offered a chain reaction of simplifications," Houbolt recalled later: "development, testing, manufacturing, launch and flight operations. All would be simplified." When the idea had first come to him, his own

reaction had been, "Oh my God, this is it. This is fantastic! If there is any idea we must push, it is this one."

And push he did, for many lonely months. Though he appeared before every NASA committee that would hear him, his efforts seemed to be in vain. Finally, late in 1961, he wrote to NASA Associate Administrator Robert Seamans, pouring out both his theory and his frustration "somewhat as a voice in the wilderness." Houbolt ended with a bold promise: "Give us the go-ahead," his letter wrote, "and we will put men on the moon in very short order—and we don't need any Houston empire to do it."

The letter scored a direct hit, persuading not only Seamans but many others among the top NASA brass. Meanwhile, LOR had begun to win over its early opponents—including Faget, who became a powerful ally. When von Braun was converted in 1962, said Houbolt, "the last hurdle had been cleared." The idea had triumphed. All that remained was to put it into effect.

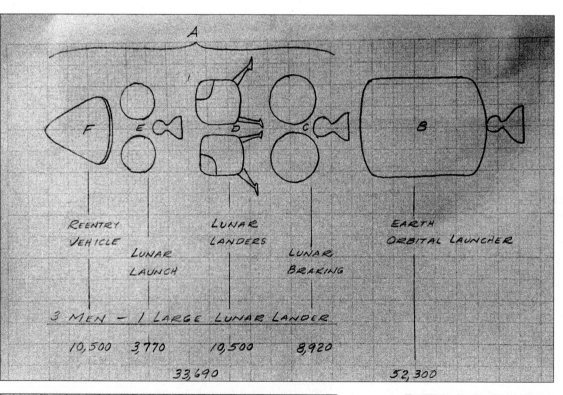

The 1961 drawing above by John Houbolt outlines one idea for an LOR spacecraft. The command module *(left)* and launch vehicle *(right)* bracket two lunar landers—having two would permit rescue if one crashed. The engine at *C* would fire to enter lunar orbit, the one at *E* to escape lunar orbit. The numbers at bottom show each segment's estimated weight.

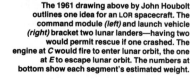

In one of Houbolt's first rough sketches, the capsule and lunar module (LM) enter a clockwise lunar orbit at ΔV_1 (the symbol Δ V stands for "change in velocity"). The LM separates from the capsule and begins its descent at ΔV_d, then lifts off to rendezvous with the capsule at ΔV_a. The spacecraft finally breaks out of its lunar orbit for the return voyage at ΔV_2.

This small wooden model of the lunar module (LM) stands on five paper-clip legs. The knobs on the upper section—the LM's manned stage—represent docking ports leading to the Apollo spacecraft. The model was built in 1962 by Grumman Aircraft, which later obtained the $1.61 billion contract to build 15 LM flight vehicles, 10 test vehicles and two simulators.

John Houbolt displays his original LOR proposal in early 1969—eight years after first presenting it and less than half a year before the first manned moon landing. NASA gave Houbolt its Exceptional Scientific Achievement Award for his "foresight and perseverance."

The launch vehicle and spacecraft designed to implement the lunar-orbit rendezvous moon shot may have been smaller and simpler than those required by either earth-orbit rendezvous or direct ascent—but they were monsters of bulk and complexity nonetheless. The assembled behemoth *(opposite)* stood 364 feet tall and weighed more than six million pounds. The three-stage Saturn V (which was responsible for almost all of the bulk) was the most powerful rocket ever built, providing 7.5 million pounds of thrust at launch.

Atop the Saturn booster sat the two-part command and service module (CSM). The command module (CM), which measured 13 feet across the base, was the only part of the assembly that returned to earth. It weighed about 13,000 pounds at lift-off and served as both command center and living quarters. The astronauts sat three abreast, with the CM pilot at the center, facing the controls in the capsule's nose. There were sleeping bags mounted under the left- and right-hand couches.

The cylindrical service module (SM) was attached to the CM like an oversized lunar backpack. It carried the oxygen supply for the ship, as well as the engine that powered the capsule into and out of lunar orbit. The service module also carried the small reaction-control thrusters that were used to adjust the spacecraft's attitude; when set in the "barbecue mode," these small rockets kept the craft slowly rotating, as though on a spit, evenly distributing the extreme temperatures of space.

The lunar module (LM) was the only one of Apollo's three main parts that actually landed on the moon. It was also the world's first true spacecraft—the first vehicle that was created to fly exclusively in the vacuum of outer space. Thus aerodynamics and much else could be ignored in its design. The squat, functional LM—given the apt code name *Spider* for its first manned test—spent blast-off protectively cradled between the rocket and the CSM because its paper-thin aluminized Mylar skin would have broken up or burned away in the high-speed rush through the earth's atmosphere. Attached to one of the forward legs of the LM was a flimsy ladder, created for use on the moon's surface, that would have crumpled had the astronauts tried to use it on earth.

The LM went through several versions as engineers struggled to expand its capacity and reduce its weight. In its final form, the lunar module weighed 16 tons and contained 30 miles of wiring, eight radio systems and 15 antennas. More than two thirds of the total weight was in the LM's lower (or descent) stage, which housed the descent rocket, fuel and water tanks, and equipment—such as the electric Lunar Roving Vehicle, or Rover—for exploring the moon's surface.

The descent stage also served as a launch pad for the ascent stage, the cramped portion that held the craft's flight computer and cockpit. Here the astronauts stood anchored in their stations by armrests, pulleys and Velcro strips attached to the floor. The ascent engine provided 3,500 pounds of thrust for the return to the CSM, while 16 small rockets on outriggers around the LM enabled the craft to maneuver. The LM's flanks and three windows (one for each astronaut and an overhead window for docking) were shielded against both micrometeoroids and the sun. There was radar for both landing and rendezvous. The ascent stage also had two hatches, one at the top for docking, another in the side. Through the side hatch would pass the first man to touch the moon.

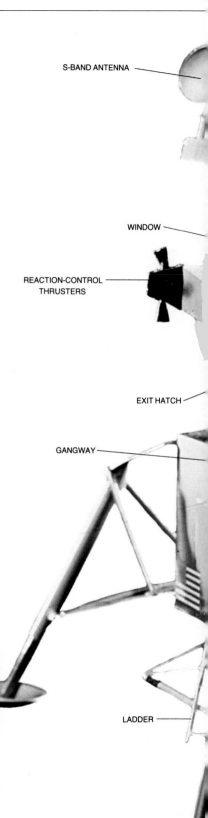

S-BAND ANTENNA

WINDOW

REACTION-CONTROL THRUSTERS

EXIT HATCH

GANGWAY

LADDER

This 1962 model of the lunar module (LM) is a refinement of Grumman's original wooden model *(page 129)*. It too has five legs and two docking ports; but four windows, antennas and capsule-shaped fuel tanks have been added.

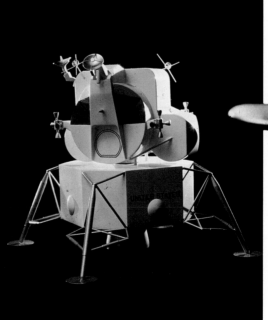

A 1965 LM model shares little more than its buglike outline with its predecessor. The front docking port has become an exit hatch, the body has been pared down, fuel tanks reshaped and two windows and a leg removed.

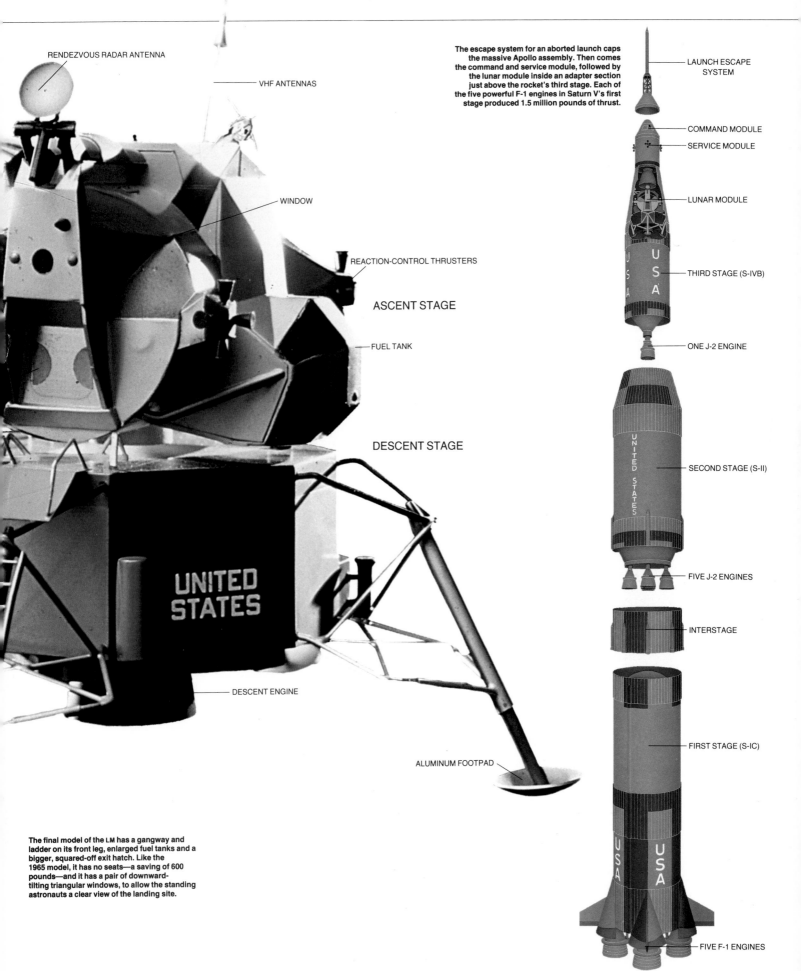

RENDEZVOUS RADAR ANTENNA

VHF ANTENNAS

WINDOW

REACTION-CONTROL THRUSTERS

ASCENT STAGE

FUEL TANK

DESCENT STAGE

DESCENT ENGINE

ALUMINUM FOOTPAD

UNITED STATES

The escape system for an aborted launch caps the massive Apollo assembly. Then comes the command and service module, followed by the lunar module inside an adapter section just above the rocket's third stage. Each of the five powerful F-1 engines in Saturn V's first stage produced 1.5 million pounds of thrust.

LAUNCH ESCAPE SYSTEM

COMMAND MODULE

SERVICE MODULE

LUNAR MODULE

THIRD STAGE (S-IVB)

ONE J-2 ENGINE

SECOND STAGE (S-II)

FIVE J-2 ENGINES

INTERSTAGE

FIRST STAGE (S-IC)

FIVE F-1 ENGINES

The final model of the LM has a gangway and ladder on its front leg, enlarged fuel tanks and a bigger, squared-off exit hatch. Like the 1965 model, it has no seats—a saving of 600 pounds—and it has a pair of downward-tilting triangular windows, to allow the standing astronauts a clear view of the landing site.

131

The flight begins, as always, with ignition: Wrapped in a mantle of flame, the Saturn V climbs free of the tower, of the earth and of gravity itself. The first stage, its energy spent, falls to the sea; the next stage kicks in and then falls away in turn. The last stage ignites briefly, establishing earth orbit, then reignites to break the orbital grip and hurtle the craft on its 240,000-mile journey to the moon.

En route, the command and service module (CSM) separates from the rocket, turns around and plucks the lunar module (LM) from the rocket's shell. As the composite craft goes behind the moon, the service module engine slows it to slip into lunar orbit. While the command module pilot remains behind, the mission commander and lunar module pilot squeeze into the LM, detach it from the CSM and fire the descent rocket, gently lowering the craft. The LM hovers briefly at 700 feet while the astronauts study the landing site, then stops again just above the surface; the probes on the legs make contact and the LM settles onto the moon.

The return voyage is relatively simple. The ascent stage of the LM blasts off *(photographs, right)* from the descent stage, which serves as the launching pad. Thus reduced, the LM achieves rendezvous and docks with the orbiting CSM. The astronauts crawl back into the CSM, the LM is jettisoned and the service module rocket blasts the CSM free of lunar orbit. Just before reentry into the earth's atmosphere, the command module jettisons the service module. Then, a mere fraction of what left the earth eight days earlier, the craft begins its descent and splashes down. □

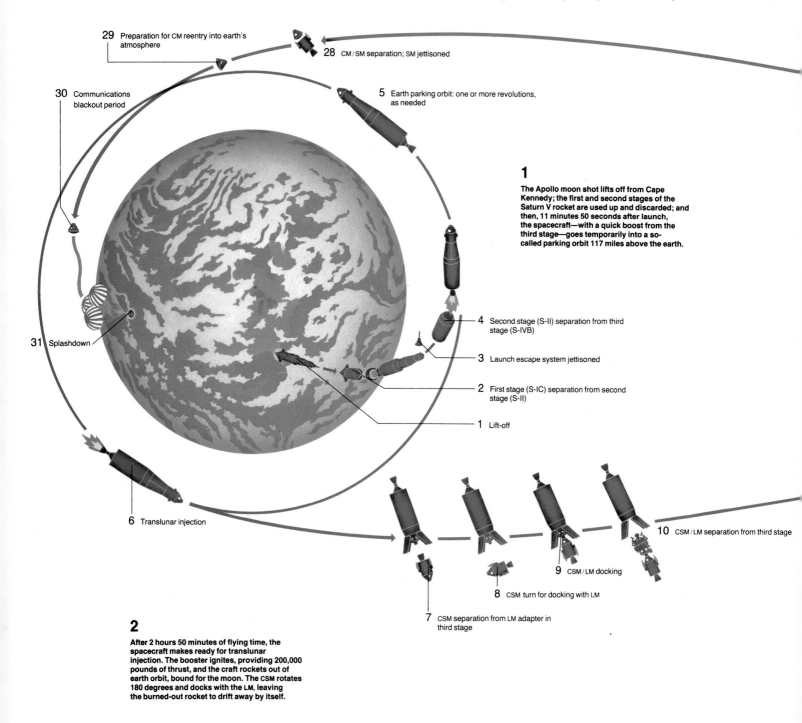

29 Preparation for CM reentry into earth's atmosphere

28 CM / SM separation; SM jettisoned

30 Communications blackout period

5 Earth parking orbit: one or more revolutions, as needed

1

The Apollo moon shot lifts off from Cape Kennedy; the first and second stages of the Saturn V rocket are used up and discarded; and then, 11 minutes 50 seconds after launch, the spacecraft—with a quick boost from the third stage—goes temporarily into a so-called parking orbit 117 miles above the earth.

4 Second stage (S-II) separation from third stage (S-IVB)

3 Launch escape system jettisoned

31 Splashdown

2 First stage (S-IC) separation from second stage (S-II)

1 Lift-off

6 Translunar injection

10 CSM / LM separation from third stage

9 CSM / LM docking

8 CSM turn for docking with LM

7 CSM separation from LM adapter in third stage

2

After 2 hours 50 minutes of flying time, the spacecraft makes ready for translunar injection. The booster ignites, providing 200,000 pounds of thrust, and the craft rockets out of earth orbit, bound for the moon. The CSM rotates 180 degrees and docks with the LM, leaving the burned-out rocket to drift away by itself.

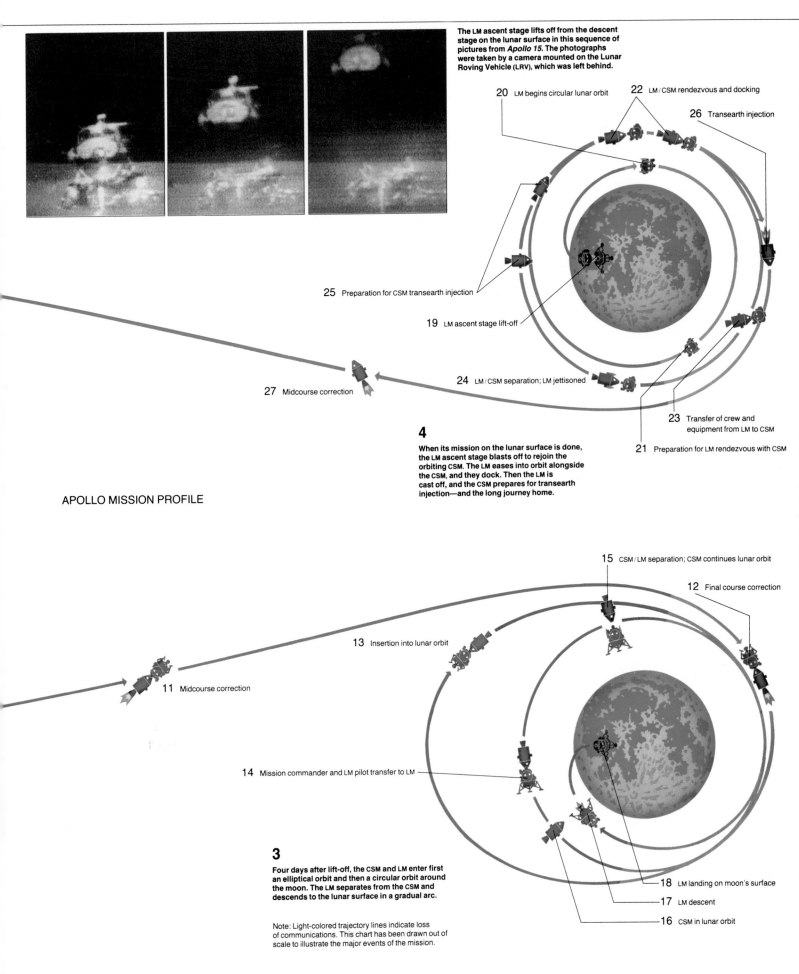

The LM ascent stage lifts off from the descent stage on the lunar surface in this sequence of pictures from *Apollo 15*. The photographs were taken by a camera mounted on the Lunar Roving Vehicle (LRV), which was left behind.

20 LM begins circular lunar orbit

22 LM/CSM rendezvous and docking

26 Transearth injection

25 Preparation for CSM transearth injection

19 LM ascent stage lift-off

24 LM/CSM separation; LM jettisoned

27 Midcourse correction

23 Transfer of crew and equipment from LM to CSM

21 Preparation for LM rendezvous with CSM

4

When its mission on the lunar surface is done, the LM ascent stage blasts off to rejoin the orbiting CSM. The LM eases into orbit alongside the CSM, and they dock. Then the LM is cast off, and the CSM prepares for transearth injection—and the long journey home.

APOLLO MISSION PROFILE

15 CSM/LM separation; CSM continues lunar orbit

12 Final course correction

13 Insertion into lunar orbit

11 Midcourse correction

14 Mission commander and LM pilot transfer to LM

18 LM landing on moon's surface

17 LM descent

16 CSM in lunar orbit

3

Four days after lift-off, the CSM and LM enter first an elliptical orbit and then a circular orbit around the moon. The LM separates from the CSM and descends to the lunar surface in a gradual arc.

Note: Light-colored trajectory lines indicate loss of communications. This chart has been drawn out of scale to illustrate the major events of the mission.

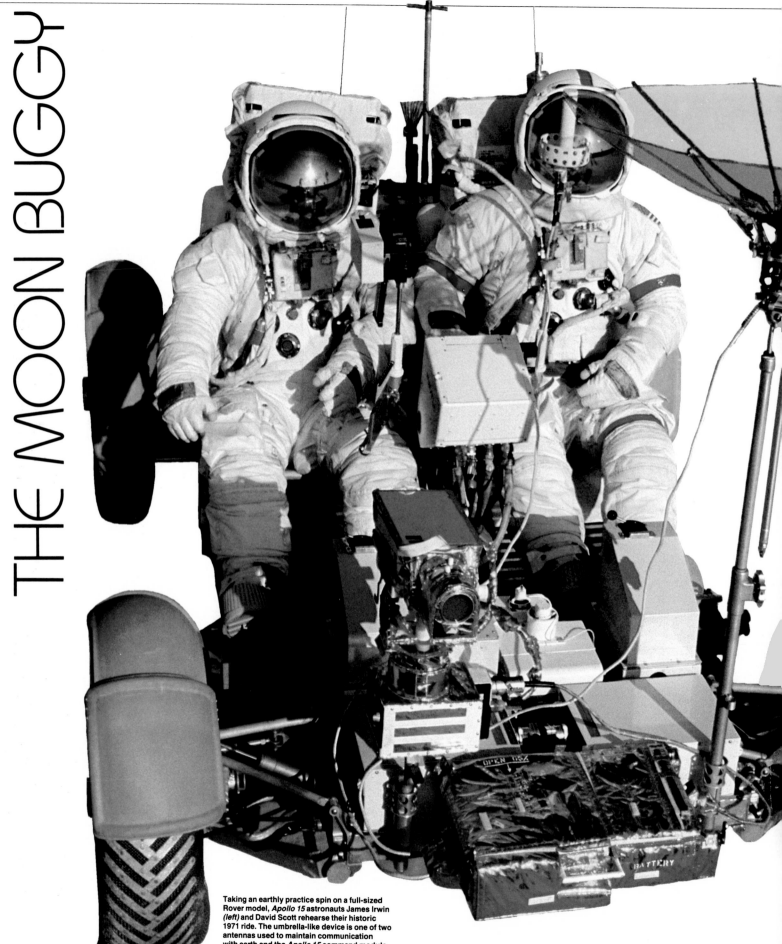

THE MOON BUGGY

Taking an earthly practice spin on a full-sized Rover model, *Apollo 15* astronauts James Irwin *(left)* and David Scott rehearse their historic 1971 ride. The umbrella-like device is one of two antennas used to maintain communication with earth and the *Apollo 15* command module.

Although it looked like a stripped-down dune buggy slapped together from odds and ends, the first automobile on the moon was one of the most unusual and expensive cars ever built. Designated the Lunar Roving Vehicle and nicknamed Rover, it was one of three driven on the moon during the Apollo program.

Rover had no steering wheel and no brake pedal. A single control for stopping, starting and steering was between the seats and could be used by either passenger. Small 1/4-horsepower electric motors in each wheel hub—each pair powered by two 36-volt batteries—produced a top speed of scarcely 10 miles an hour. But from the remote-control color television camera at the front to the teflon moon-rock collecting bags at the rear, the little vehicle was superbly equipped for its brief career.

Oversized tires of flexible wire mesh were designed to carry Rover and its two astronaut-explorers across crevasses more than two feet wide and up slopes as steep as 25 degrees. And when the day's work was done, Rover's computerized navigation system guided it unerringly back across the uncharted landscape to its lunar-module base. □

A wide-track Rover tire, with a titanium tread fastened to its layer of wire mesh, dwarfs a conventional automobile tire. Built to cope with the dusty, rugged lunar surface, Rover's tires had a life expectancy of 112 miles.

Suspended to simulate the effect of the moon's reduced gravity, a Rover model is driven over a sandy, rock-strewn track by astronaut Scott *(right)*. In such tests, the car tended to fishtail at high speed—10 miles per hour.

Rover's navigation panel, mounted above its single control lever, contains a small computer and a gyroscope, essentials of an inertial guidance system that automatically calculates the direction and distance to the starting point without a compass or radio signal.

LUNAR ODDITIES

In the 1960s, human inventiveness was put to the test of creating unique equipment for moon exploration, and the results were often startling. Made for a dusty, airless environment where gravity is only a sixth of that on earth and where daily temperatures range from 250° F. to –250° F., early moonsuits and lunar vehicles resembled strange, unworldly toys.

A robot-like moonsuit *(right)*, equipped with food, water, oxygen, air conditioning and even a tiny stove, eventually proved too cumbersome for the astronauts. Early moon-vehicle prototypes similar to those below, fitted with flexible metal wheels for traversing the rock and sand of the moon's surface, evolved into the Lunar Roving Vehicle *(pages 134-135).*

Looking like part of an oldtime circus wagon, a full-sized moon-vehicle model demonstrates its collapsible metal wheels in 1967. Each wheel was powered by an electric motor in its hub, a technique retained in the final version.

Tucked inside the pressurized cylinder of an experimental moonsuit, with his arms and legs extending through the accordion-pleated appendages, inventor Allyn Hazard tests his creation in California's Mojave Desert.

Another proposed vehicle, with living quarters for two astronauts, moves on pinwheel-shaped metal wheels in a 1967 test drive.

apollo

MISSION PORTFOLIO

Astronauts Gus Grissom, Ed White and Roger Chaffee were undergoing a routine pad test on January 27, 1967, just weeks before what was to be the first manned Apollo launch. Strapped securely into their flight seats and sealed off from the outside, they were unprepared for the sudden fire that engulfed them. In seconds they were dead. What had started the blaze was never precisely determined (an electrical spark may have been the cause), but the cockpit's pure-oxygen atmosphere, pressurized to 16 pounds per square inch, had created a hazardous environment; even normally flame-retardant materials, including the spacesuits, had burned furiously. A shocked nation mourned the loss of veterans Grissom and White and of 31-year-old Chaffee, the youngest astronaut assigned to a space flight. The spacecraft was redesigned and its oxygen atmosphere tempered with nitrogen while the craft was on the ground, but the tragedy set the Apollo program back more than a year and a half.

Reviewing their flight plans at a private home in Hollywood Hills earlier in the winter, Chaffee *(left),* White and Grissom squint and shade their eyes in the California sunshine. "There's always a possibility of catastrophic failure," Grissom had said once. "You just plan as best you can to take care of all these eventualities, and you get a well-trained crew and you go fly."

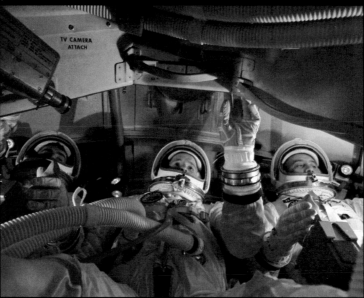

In an early training session, White, Chaffee and Grissom sit helmeted, fully suited up and connected to oxygen hoses in the Apollo Mission Simulator, a replica of the spacecraft they were testing on the day of the fire.

The charred inside of the destroyed craft is stark evidence of the fire's destruction. Oddly, while virtually everything but metal was badly burned, a portion of the flight plan *(lower right)* survived with only a few pages singed.

In a solemn procession, Air Force Lieutenant Colonel Gus Grissom's casket is borne along to its final resting place in Arlington National Cemetery. Navy Lieutenant Commander Roger Chaffee was also buried at Arlington, and Air Force Lieutenant Colonel Ed White at his alma mater, West Point Military Academy.

As "Taps" is sounded, Grissom is honored by his fellow astronauts Glenn, Shepard, Cooper, Carpenter and Young. All except Young had been with Grissom among the seven originally chosen; Young had been Grissom's partner during the flight of *Gemini 3.*

Betty Grissom watches as the flag from her husband's coffin is folded in the military manner. As an astronaut's widow, Betty Grissom got no special compensation from the government for her husband's death. After suing the builder of the spacecraft, she later settled out of court for $350,000 for herself and her sons.

APOLLO 7

The launch of *Apollo 7* on October 11, 1968, signaled the resumption of manned U.S. space flights after a hiatus of more than 20 months. For almost 11 days, Wally Schirra, Donn Eisele and Walt Cunningham put the command and service module (CSM) through its paces in earth orbit, delighting in its roominess. "We actually had living quarters," Cunningham said later, "not just a place to sit." One of their major objectives was to rendezvous—without benefit of radar—with the second stage of their Saturn booster. In the last mile, closing maneuvers were made by eyeballing the target. "That rendezvous was a nightmare," commander Schirra said later. But the veteran pilot's hand at the controls never faltered. "We just slid right up the pipe and onto the target," said navigator Eisele. "It was a great feeling."

Apollo 7 seems to brush past the 525-foot-high building where the Saturn V rockets were assembled. Taken from an airplane at 35,000 feet, the photograph is an illusion; the building is about five miles from the launch site.

Spewing flames, Apollo 7 gains speed as it climbs into the Florida sky. All three crew members professed excitement at takeoff, but the pulse of veteran Schirra (the first and only astronaut to pilot Mercury, Gemini and Apollo spacecraft) registered only 87.

The fiery moment of first-stage separation comes 39 miles up, at 2 minutes 45 seconds into the flight. A few seconds later, the next stage ignited to drive the spacecraft into orbit.

A tearful Harriet Eisele grips her four-year-old son, Jon, as they watch the lift-off on television at home in Houston. She felt sure that the rocket had blown up. Days later, as splashdown approached, she had to fight against breaking into sobs. "It's agony," she said.

The spacecraft launch adapter on the rocket's second stage opens like a giant flower during *Apollo 7's* simulated docking high over the Gulf of Mexico. Close-in maneuvering with the spacecraft, Eisele said later, was "rather like one car overtaking another, but a car with very weak brakes and not much acceleration."

Etched with remarkable clarity, the Himalayas pass beneath the high-flying *Apollo 7.* With at least one window in the spacecraft always facing such dazzling sights, Cunningham grew a bit blasé. On the 11th or 12th pass he found himself thinking, "Oh, it's the Himalayas again."

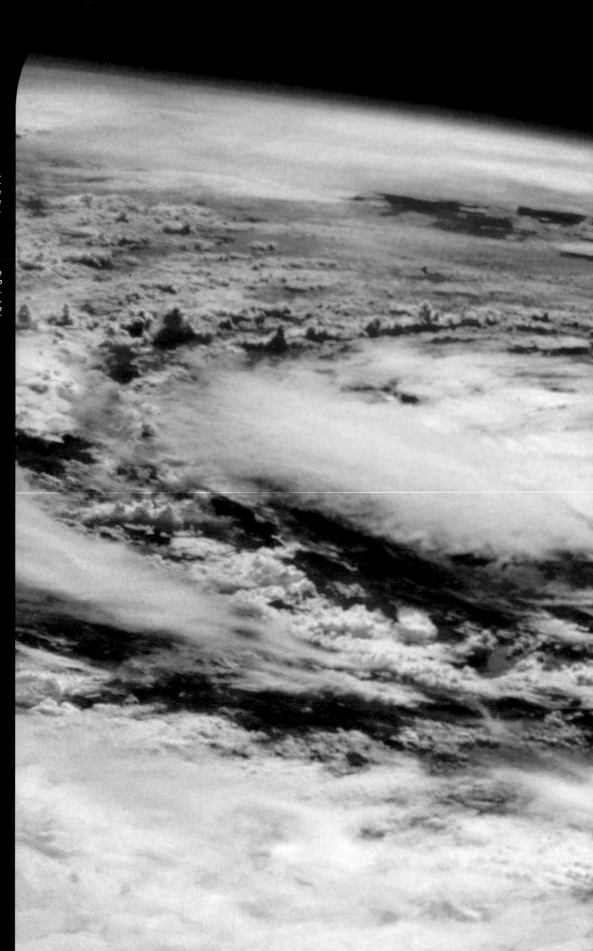

Seething and swirling below the spacecraft, Hurricane Gladys reveals its fury in a 200-mile-wide swath off the Cuban coast. The crew also viewed an immense typhoon north of the Philippines; its eye alone was 70 miles across.

Deep into the mission, astronauts Schirra, Eisele and Cunningham *(inset opposite, left to right)* appear to be rather the worse for wear. Their "grungy" state, Schirra commented later, was a far cry from the popular notion that "we're up here dashing around as space heroes."

Until Apollo 8, no one had ventured beyond earth orbit. But the team of Frank Borman, Jim Lovell and Bill Anders flew all the way to the moon, which they circled 10 times before returning to earth. To achieve this milestone required tremendous propulsion, supplied by a massive rocket: the Saturn V. So immense that its sections had to be transported by water on special barges, the rocket took five years to build and absorbed the know-how and energies of 325,000 people in 12,000 firms. As tall as a 36-story building, it had 11 engines; the five engines that powered the first stage alone were designed to develop 160 million horsepower, the capacity of 86 Hoover Dams. On launch day—December 21, 1968—this awesome device functioned without a hitch.

Surrounded by work platforms *(right),* the huge rocket is assembled at Cape Kennedy. Each stage was stacked in position by computer-assisted cranes powerful enough to lift hundreds of tons—and precise enough to touch that weight to an egg without breaking it.

Anders, Lovell and Borman *(above, left to right)* rehearse launch procedures in a replica of their command module. "I have no hesitancy about the hardware," asserted Borman. He was a Gemini veteran, as was Lovell; Anders was the only one who was new to space travel.

Inside the first stage, in reality a cavernous fuel tank, workers check antislosh rings. Before being fueled for the mission, each section was meticulously washed. Even a thumbprint could leave enough grease to react with the liquid-oxygen oxidizer and cause an explosion.

Saturn's five huge first-stage engines, which will lift *Apollo 8* off the pad, are adjusted by workers at the California plant. Each of the engines, as big as a two-and-a-half-ton

As *Apollo 8* soars upward, a bird flaps by with seeming unconcern. More than 50,000 acres at the Cape had been set aside as a wildlife refuge, and every blast-off would send birds by the thousands fluttering into the air. But eventually they would return to their habitats.

Safely back on earth, Frank Borman—with Bill Anders and Jim Lovell looking on—thanks the aircraft carrier crew for their assistance in the recovery. He also apologized for having disrupted their Christmas holidays.

Looking across the bleak moonscape, the astronauts behold the blue-white sphere of earth. The line separating day and night bisects Africa. Lovell imagined himself a traveler from another planet, wondering if earth was inhabited.

The purpose of the 10-day *Apollo 9* mission was to flight-test the ungainly-looking lunar module (LM) in earth orbit. A vehicle intended solely for use in space, the LM was so frail that its flanks would crumple if subjected to flight in earth's lower atmosphere. After lift-off on March 3, 1969, the crew of Jim McDivitt, Dave Scott and Russell Schweickart performed for the first time the tricky maneuver of withdrawing the LM from the adapter on the Saturn booster's third stage, where it had been sheltered during launch. Later, McDivitt and Schweickart crawled into the LM—code-named *Spider*—fired it up and flew off into space on their own. After eight hours—during which they changed orbits and let as much as 100 miles separate them from Scott in the command and service module—the two astronauts successfully rendezvoused and docked with the mother ship. The LM—which had no heat shield to protect it— was jettisoned before reentry. Despite its appearance, the gawky LM had proved itself ready for the lunar job ahead.

Schweickart *(left)*, McDivitt *(seated)* and Scott pause during a training session for a portrait in full pressure suits. Schweickart was the rookie; McDivitt had been command pilot of *Gemini 4*, and Scott had been aboard *Gemini 8* when it made the first docking in space.

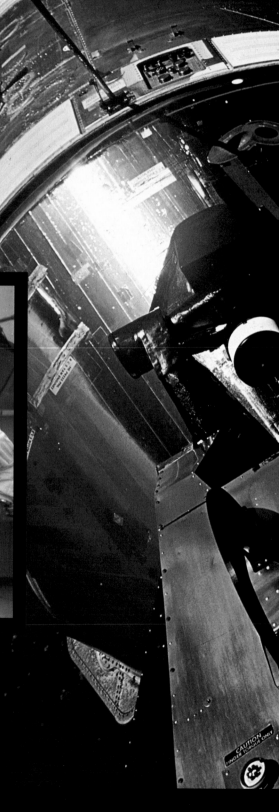

Far above the blue earth on the fourth day of the mission, Scott stands in the command module's open hatch and gazes toward the as-yet-untested LM, now mated to the command module. The photograph was taken by Schweickart, who had emerged from the craft earlier for an EVA to test the lunar spacesuit.

The folded-up lunar module (LM) rides within its protective shroud as it did during launch and for the first orbit around the earth. This photograph was taken just as the command and service module prepared to dock with the LM in order to withdraw it from its shell.

Its propeller-like antenna jutting toward earth, the command and service module passes over Hawaii in a picture taken from the LM during its first manned test flight. Scott flew by himself in the command module for a six-hour period while the two craft were separated.

Free and upside down in the black sky of
space, the lunar module is put through its paces
on the fifth day of the *Apollo 9* mission
by Schweickart and McDivitt. The cloud-covered

With earth clouds and the LM reflected in his
visor, Schweickart pauses during a space walk to
take photographs from the LM's "front
porch." One photograph he took was of Scott
(page 153), who then snapped this picture.

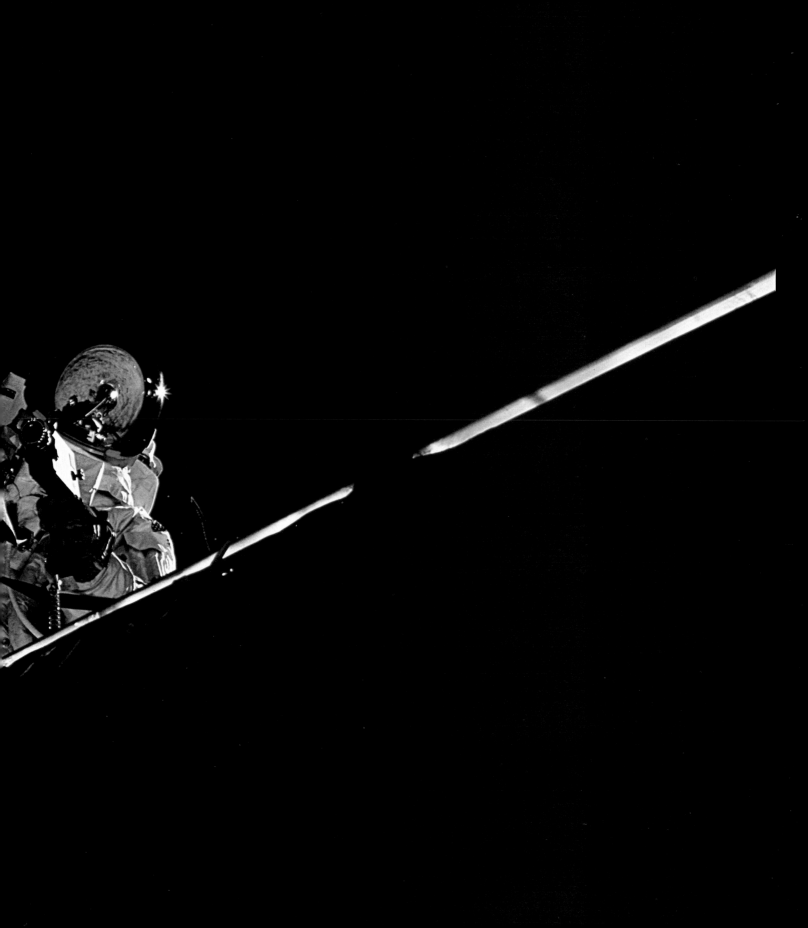

One of the three gigantic parachutes that controlled the descent of the spacecraft billows against the sky as *Apollo 9* heads for splashdown. The chutes, packed tightly atop the command module during the mission, slowed the spacecraft's fall to a gentle 22 mph.

Back in Houston, the astronauts' wives celebrate splashdown with champagne—and matching jerseys that jokingly denote their titles: Clare Schweickart (Lunar Module Pilot Wife), Lurton Scott (Command Module Pilot Wife) and Pat McDivitt (Commander Wife).

An exultant welcoming party greets the three
astronauts on their arrival back in Houston.
Scott's eight-year-old daughter, Tracy *(center)*,
showed her father a theme she had written
in his absence. In it she described going to the
moon with her parents and brother. It was dusty
there, and instead of walking they bounced. The[n]
they went home. "It was nice to be home,"
she wrote in conclusion, "and we decided that
next Saturday we would go to the zoo."

By the time it splashed down on May 26, 1969, *Apollo 10* had virtually assured NASA that the goal of a lunar landing before 1970 could be met. On the fourth day of their eight-day mission, Tom Stafford and Gene Cernan took the lunar module *Snoopy* to within nine miles of the lunar surface, leaving John Young in the command module *Charlie Brown* to send back the first color TV pictures of the moon. *Snoopy's* descent radar performed superbly; but there was one moment of terror later, when *Snoopy* jettisoned its descent stage to simulate lunar lift-off. A switch was in the wrong position, and suddenly *Snoopy* began gyrating violently. Stafford wrestled it back under control, but the astronauts' earthy expletives caused a minor flap back home: A minister complained to NASA that the crew had carried "the language of the street" to the moon.

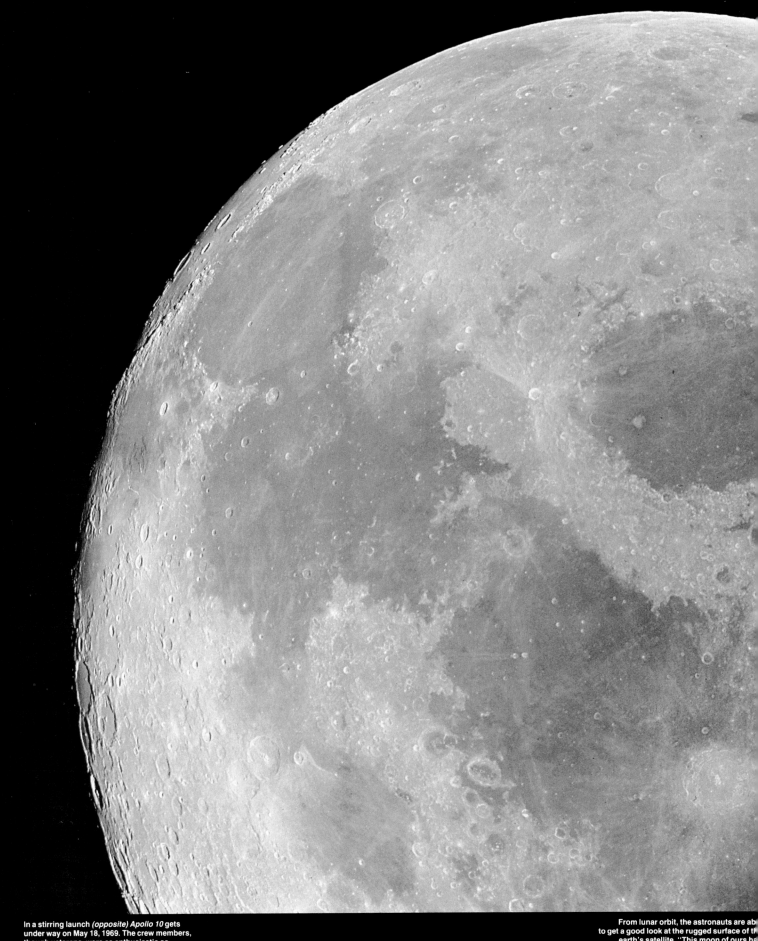

In a stirring launch *(opposite) Apollo 10* gets
under way on May 18, 1969. The crew members,
though veterans, were as enthusiastic as
rookies. "What a ride!" Gene Cernan exclaimed

From lunar orbit, the astronauts are ab
to get a good look at the rugged surface of th
earth's satellite. "This moon of ours ha
a rough beginning somewhere back there," on

The LM moved near the CSM before its dress rehearsal of a lunar descent. At 65,000 feet the LM's radar sensed the lunar surface and began gathering data on altitude and descent rate.

the far side of the moon was photographed from *Apollo 10*. Movies and still photographs brought back by the crew helped NASA zero in on landing sites for future missions.

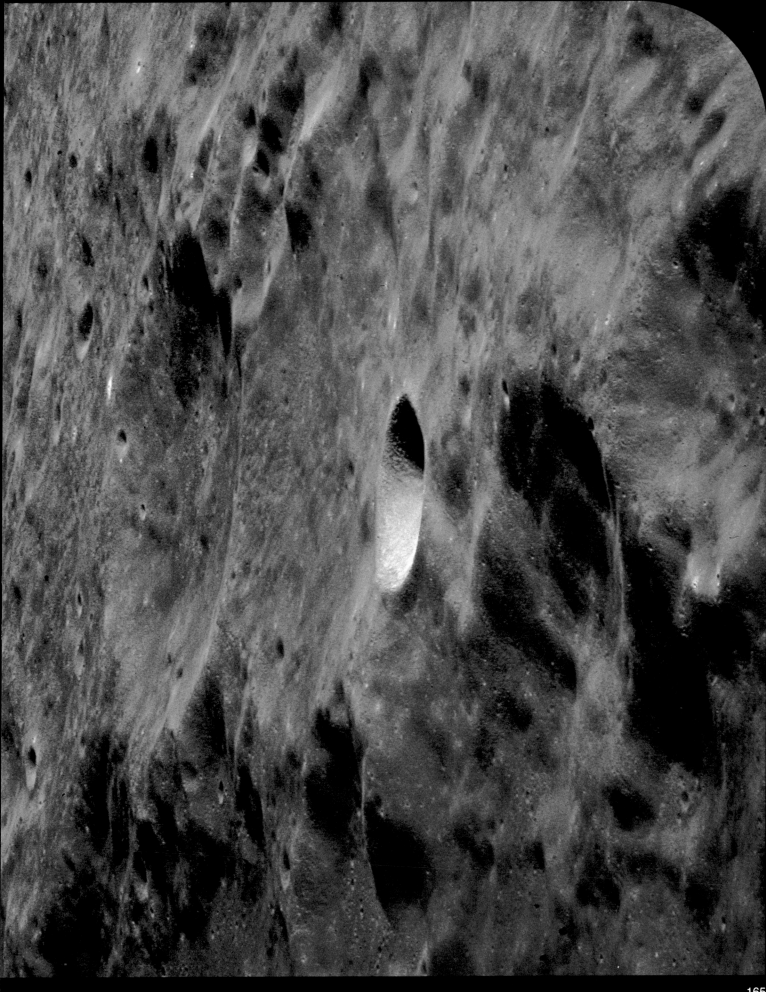

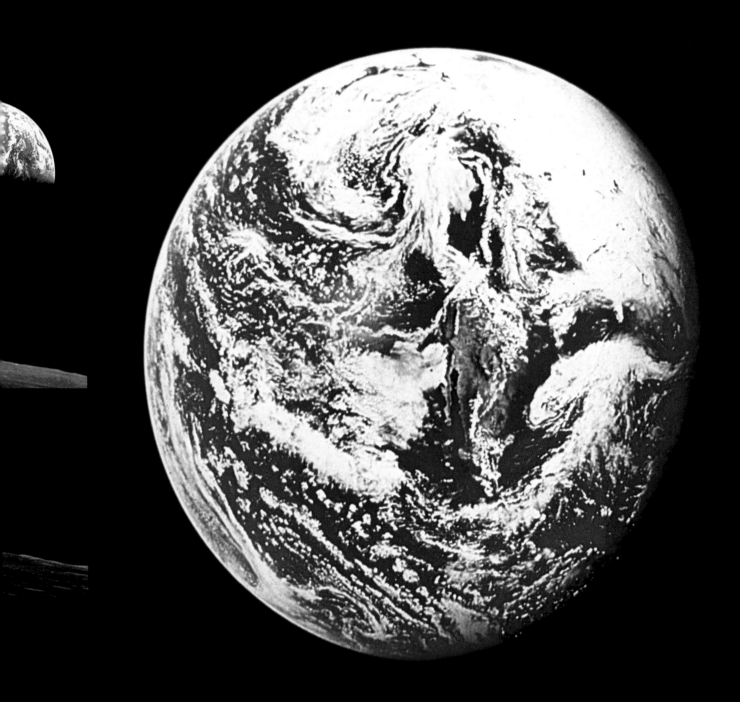

A fully risen earth floats gloriously in the
blackness of space. The Baja Peninsula and the
west coast of Mexico are clearly visible
through swirling cloud formations *(center)*.

A trio of billowing chutes gently lowers the
Apollo 10 command module to a dawn
splashdown in the Pacific. Recovery this time
, was swift: The spacecraft's landing was
only three miles from the carrier *Princeton*

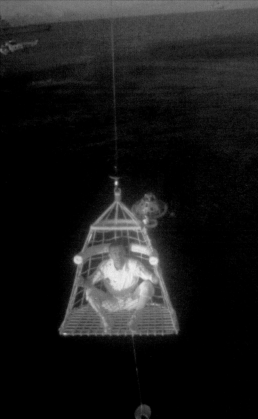

A cagelike rescue lift hoists Gene Cernan up to the recovery helicopter. Behind him is the command module, riding the ocean swells in its buoyant flotation collar.

Faye Stafford, Barbara Young and Barbara Cernan *(below, left to right)* share their relief and jubilation in Houston. "One, two, three!" Faye Stafford exclaimed, watching the chutes open on TV. "Nice to be counting frontwards again."

Cernan, Young and Stafford don the traditional baseball caps for a welcoming ceremony aboard the *Princeton*. In Houston NASA's Thomas Paine announced, "Today, this moment, we know we can go to the moon. We will go to the moon."

On July 20, 1969, the world listened spellbound in front of televisions and radios as an American astronaut set a tentative foot on the moon. "That's one small step for a man," said Neil Armstrong, "one giant leap for mankind." *Apollo 11* was the pay-off—and it was glorious. Armstrong and teammate Buzz Aldrin moved cautiously on the powdery surface at first, setting up experiments and collecting soil and rock samples. Then, gaining confidence, they hopped like kangaroos in the weak lunar gravity. Armstrong took so many pictures that Houston had to remind him four times to get on with his other tasks. After 21½ hours on the surface, the lunar module *Eagle* rose to rejoin pilot Mike Collins in the orbiting command module *Columbia*. But where the *Eagle* had landed, the first human footprints were left behind in the lunar dust.

Bathed in floodlights, *Apollo 11* stands moored to its gantry at Cape Kennedy, awaiting final preparations and its three-man crew. A crescent moon—the spacecraft's distant destination—hangs in the night sky.

Topped by the launch escape tower, *Apollo 11* begins its ponderous lift-off *(bottom left)* at 9:32 a.m. on July 16, 1969. Seconds later, as the Saturn booster moves past the gantry *(left)*, the rocket shakes loose half a ton of ice that had formed around the second-stage tank holding the supercold liquid propellant.

Throngs of journalists record the historic launch from a position three miles distant. For fifteen seconds after the ignition, their cameras were recording what seemed to be a silent sight. Then the delayed sound arrived, building to an earth-shaking roar.

Former President Lyndon Johnson and Lady Bird Johnson watch delightedly as the rocket begins its lift-off. As it climbed higher, Johnson craned his neck to follow it while the former First Lady wiped away a tear.

granular surface, Aldrin poses for a lunar portrait. The reflections in his visor show all the components of this first visit to the moon: scientific equipment, the LM and astronaut-photographer Armstrong, who took the picture.

On the moon at last, Aldrin sets up an experiment sponsored by the Swiss government to study the solar wind. Deployed for an hour and 17 minutes, the aluminum sheet caught 10 trillion atoms of chemical elements streaming from the sun at supersonic speeds.

Next, Aldrin arranges seismic equipment to record lunar tremors. Left in place after the astronauts' departure, it radioed evidence of a moonquake and other minor shakes believed to be caused by landslides in nearby craters.

Aboard the carrier *Hornet* after splashdown, the astronauts wear BIGs—biological isolation garments—as they head for the quarantine trailer. They remained in quarantine for 18 days while scientists investigated their possible contamination from moon microbes.

Quarantined in the trailer, the lunar voyagers grin at the *Hornet's* admiring crew. After thorough examination, the astronauts were pronounced uncontaminated—as were the soil samples they had brought back—and the elaborate quarantine measures were halted after *Apollo 14.*

With the moon landing accomplished, the thrust of the Apollo program shifted from pioneering space travel to scientific inquiry. In November 1969, while Dick Gordon orbited overhead in the Apollo 12 command module *Yankee Clipper,* Pete Conrad and Alan Bean faultlessly maneuvered the lunar module *Intrepid* down to the moon's Ocean of Storms. With Conrad exclaiming, laughing and humming delightedly, the astronauts set up a complex package of experiments, powered by a small nuclear generator, to measure moonquakes, the moon's magnetic field, electromagnetic activity and other phenomena. In 7 hours 37 minutes on the surface—nearly three times as long as the time spent by the crew of Apollo 11—Conrad and Bean collected a vast store of rocks that would reveal much about lunar history. En route home, Apollo 12's astronauts were treated to a rare bonus: an eclipse of the sun by the earth, something that can be seen only from space.

Rookie Al Bean *(left)* joins veterans Dick Gordon and Pete Conrad in front of the Saturn rocket that would launch *Apollo 12* on November 14, 1969. This was Conrad's third space flight. In 1965 he had spent eight days aloft in Gemini 5; a year later he and Gordon had been partners aboard *Gemini 11.*

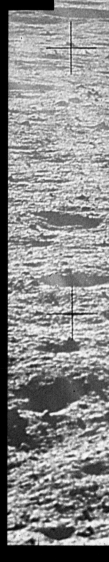

Conrad inspects Surveyor 3, a picture-taking robot that had soft-landed on the moon in April 1967. Parts of the Surveyor were brought back to Houston, where a thorough inspection proved that it had never been struck by a meteor during its lunar stay. On the horizon 600 feet away sits the lunar module *Intrepid.*

Kicking up lunar dust with each step, Al Bean strides across the moon with a package of instruments to be set up. Although the package weighed 190 pounds on earth, lunar gravity made it a manageable 30 pounds. In the background is a dish antenna for communicating with earth.

appears motionless in orbit as it waits for the
returning lunar module. The propeller-like
device jutting from its side is an antenna for
communication with Mission Control.

One frame *(above)* from a movie that was made
by the astronauts during the flight of
Apollo 12 back home captures a virtually total
eclipse of the sun by the earth.

A low sun gives an auburn hue to the moon
as seen from the *Intrepid*, the lunar module of
Apollo 12. Near the horizon is the crater

Apollo 12's largest rock sample *(above)* weighed less than a pound on the moon but more than five on earth. When the astronauts brought this specimen back, scientists had learned much from the rocks returned by *Apollo 11.* One *Apollo 11* sample is shown at right, photographed in polarized light to reveal its minerals: black, titanium-bearing ilmenite, clear crystals of feldspar, rainbow-hued pyroxene. From both missions, scientists found that the moon's predominant rocks are volcanic igneous, similar to those on the earth, an indication that both bodies had the same origins. Both earth and moon are some 4.6 billion years old, but many moon rocks were crystallized only 3.5 billion years ago, suggesting that a catastrophic eruption occurred when the moon was a billion years old. The rocks also showed that the moon has never had any water. Its "seas" were probably formed by the spreading of molten lava during the long-ago catastrophe.

Apollo 13, which lifted off on April 11, 1970, nearly came to grief when an oxygen tank in the service module exploded en route to the moon. As their first-person account reveals, the crew—mission commander James Lovell, command module pilot Jack Swigert and lunar module pilot Fred Haise—stayed alive through their own quick thinking and the incredible teamwork of the ground crews at Mission Control.

LOVELL: Just before the accident, Fred Haise and I had done a routine telecast from inside the lunar module *Aquarius*. We had gone into *Aquarius* a little earlier than the flight plan called for because we were anxious to check the pressure on a helium tank. It was all right. Then we did our housekeeping chores, and while Fred worked in the LM I did the camera work and sent the pictures back to Houston. I crawled back into the command module *Odyssey*, followed by Fred, who was to close the hatch. We were 55 hours and 55 minutes into a mission that had been planned to last about 240 hours. Then we heard the bang.

HAISE: I felt the wall of the tunnel shiver. The master alarm squealed through my earphones. Almost simultaneously, Jack yelled down to me that there was a warning light on. I continued through the tunnel from the LM to the command module. By the time I got back, one main voltage reading was at the lower limit of the voltage meter. I recall a profound sinking feeling. My first thought was—well, we've blown the lunar landing. I didn't need to look at the card on the panel in front of me that listed the mission rules for an insertion into lunar orbit. I knew it was a mission rule that the loss of one electricity-producing fuel cell scrubbed a moon landing. No lunar orbit. No descent burn to the surface. No exploration of the moon. A bitter personal disappointment.

SWIGERT: The fuel cells were all going. It was only a question of time until we were without oxygen and power in the command module. I suggested that we start thinking about powering up the LM and getting its guidance system aligned. In very short order, Mission Control came back with the word that it was to be an LM lifeboat mission. We had 15 minutes before our last fuel cell went out. This was something we'd never done in training.

"I like to think that *Apollo 13* contributed to the maturing of the space program," said Jim Lovell, who logged 572 hours in space before *Apollo 13* lifted off—yet is the only astronaut to visit the moon twice without landing on it.

We'd thought about losing one or two fuel cells and an oxygen tank. But we'd never trained for losing all three cells and both oxygen tanks. If someone had thrown that at us in the simulator, we'd have said, "C'mon, you're not being realistic."

This was real. I've never seen the LM activated so fast. I could see that our last guidance fuel cell was going out, so I turned on a battery to keep the guidance system going while Jim got things lined up in the LM. Just as he got a good alignment, the last fuel cell quit. Mission Control told me to shut off the electrical power. We were going to need those batteries. I followed the steps and in a minute or two the command module was powered down. It was eerie: no lights, no radio and nothing for me to do. I drifted through the tunnel to *Aquarius* and looked at Jim and Fred. "It's up to you now," I told them.

LOVELL: We were not going to land on the moon, but getting home was not all that simple. We were, at the time it happened, 205,000 miles from earth, on a trajectory that would take us to within 60 miles of the lunar surface but was not designed to return us home if something went wrong. So Mission Control's first request was that we do a maneuver to get us back into the free-return trajectory, which is sometimes called the slingshot. This means going around the moon once, starting at the leading edge—roughly the left-hand side of the moon as seen from earth—with the proper trajectory. The moon will then put you on a free-return coast when you come around the trailing edge after one pass on the back side. That free-return trajectory was part of the flight plan for *Apollo 8*, *10* and *11*, but beginning with *Apollo 12* we went to a hybrid trajectory necessitated by the different landing sites but giving up one safety factor. Now we had to get that safety factor back. The obvious thing to do was to fire the descent-propulsion-system engine of *Aquarius*, which had been designed to power the lunar module on its final descent to the moon. Without that burn of the DPS engine, we would go around the moon all right, but we would have wound up stranded a few days later in a weird, egg-shaped earth orbit.

HAISE: I didn't delve into this thought. Maybe my subconscious didn't care to, I don't know. Anyway, I didn't bother with it.

"My life did not flash before me," Fred Haise recalled of the first moment of shock. "You focus in on what you know and what you've been trained to do, and you do it in as cold and calculating a way as you can muster."

LOVELL: That was the worst time—those few hours after the accident. I was worried about the systems in *Aquarius*. Nothing in the lunar module had been designed for the work we were now asking it to do. So there was the big feeling of relief when that engine fired for 30 seconds to boost us up and take us around the moon at an altitude of about 130 miles instead of 60 miles. Now we had our free ticket home—if our consumables held out: oxygen, electricity, especially water. That first burn of the DPS engine put us on a course that would land us in the Indian Ocean, but that was the least of my worries at this point—the very least of my worries. I didn't know until after I got back that so many nations had volunteered to help in a recovery operation, but any old ocean would do as long as it was on the earth.

SWIGERT: We adjusted to our routine readily enough. We started referring to our spacecraft as a two-room suite. The command module was the bedroom. Whenever Mission Control would ask where so-and-so is, we'd say, "he's up in the bedroom."

LOVELL: At first we were all too keyed up to worry about rest; I slept hardly at all for the first 35 or 40 hours. Eventually it dawned on me that somehow we all had to get some sleep, and we tried to work out a watch system. We weren't very successful. Events kept upsetting it and making a sensible rotation impossible. Besides, the inside of *Odyssey* kept getting colder and colder. It eventually got down pretty close to freezing point, and it was just impossible to sleep in there. Fred and I even put on our heavy lunar boots. Jack didn't have any, so he put on extra long johns. When you were moving around the cold wasn't so bad, but when you were sitting still it was unbearable. So the three

of us spent more and more of our time together in *Aquarius*, which was designed to be flown by two men—standing up, at that. There wasn't really sleeping space for two men in there, let alone three, so we just huddled in there, trying to keep warm and doze off by turns.

HAISE: I've been a lot colder before, but I've never been so cold for so long. It probably contributed to the kidney infection that I picked up. The last 12 hours before reentry were particularly bone-chilling. During this period I had to go up into the command module. It took me four hours back inside the LM before I could stop shivering.

SWIGERT: On the way home, Mission Control gave me a procedure for getting LM electricity to run the command module. That was something that had never been done before. By following the new procedures, we got LM power into the command module. We used it to recharge the reentry batteries. After that, we knew that we had a good command module electrical system. But we still didn't dare use it until the last couple of hours. Thanks to Mission Control and the guys who worked it all out in the simulators, the procedures worked perfectly. That last morning I was back in my element. I had something to do, and every switch and circuit breaker that I turned on in *Odyssey* just made me feel that much better. I forgot about being tired and didn't even notice the cold.

Jim and Fred were the same. Our teamwork was fantastic. We were one body with three heads and six hands. As tired as we were, there was never a cross word. Everybody meshed. Everybody took his share of the load.

LOVELL: After we dropped the service module and our blessed little *Aquarius*

Safely home again, Jack Swigert could joke, "I guess I had the shortest tour as a prime crew member of any astronaut: two days and a bang!" He had substituted for Ken Mattingly, who had been exposed to measles.

[before reentering the atmosphere], the rest was routine. We came home on the same systems—the same oxygen, the same battery power—that we would have used had the rest of the mission gone as planned. It probably is just as well that we did not see the extent of the damage to the service module earlier; we were sufficiently worried without knowing that a whole side panel had been blown away (or had dropped away) and the service module's innards were just hanging out there. There is an element of frustration when I think about all the hours of training, which took us so far but not quite far enough. But I do not feel bitter about that. Those hours of training were not wasted; indeed they got us home, with massive assistance from ground control. And just by getting back under these critical circumstances we did prove something about the American capacity for accomplishment under stress: You can do it if you have to do it. That was the primary accomplishment—and the triumph—of *Apollo 13*.

At Mission Control during the crisis, a tense group of astronauts and flight controllers monitors the consoles. In profile *(center, foreground)* is Alan Shepard, who would become the commander of the next lunar mission.

To get the moon program back in gear after the near tragedy of *Apollo 13*, NASA had assistance from an old pro. The commander of *Apollo 14* was Alan Shepard, 47, the first American to go into space and a man who had recently returned to the flight line after nearly a decade of being grounded by an ear disorder (it had been corrected by surgery). While Stuart Roosa remained aloft in lunar orbit, Shepard and LM pilot Edgar Mitchell touched down on the moon on February 5, 1971, in the hilly Fra Mauro region. For nearly 10 hours they crisscrossed the site, setting up a $25-million package of scientific experiments, among them one that used explosive charges to create seismic waves to measure the moon's subsurface structure. With the help of a handcart *(opposite)*, the astronauts collected 108 pounds of rocks and soil, and almost made it to the top of 400-foot Cone Crater before running out of time. At one point, Shepard staged a prank that only an old pro could expect to get away with. Extracting two golf balls that he had smuggled in his spacesuit, he used an implement from the tool cart to execute what he called a sand-trap shot, the first golf shot ever made on the lunar surface.

Almost a decade after the first time he was photographed as an astronaut with his family *(left)*, Shepard looks virtually unchanged by the passage of time. Striking the same pose, he sits between his wife, Louise, and daughter Laura, 23 and married. In front are his niece Alice Williams *(left)* and daughter Julie, both 19. The cats have given way to dogs.

Pausing during his lunar walk with the pull cart, Shepard fits together two core tubes that he then hammered into the ground in order to sample soil layers. Despite all the things that he had previously heard from fellow astronauts, Shepard was still startled by the bleakness of the lunar scene. "It certainly is a stark place here at Fra Mauro," he said.

Zigging and zagging as the astronauts pulled
it away from the LM *(background)*, the
two-wheeled handcart left three-quarter-inch-
deep tracks in the lunar soil. The pattern
reminded the Texas-born Mitchell ''of driving
a tractor through a plowed field.'' His companion
was less lyrical. Said Shepard, ''Nothing
like being up to your armpits in lunar dust.''

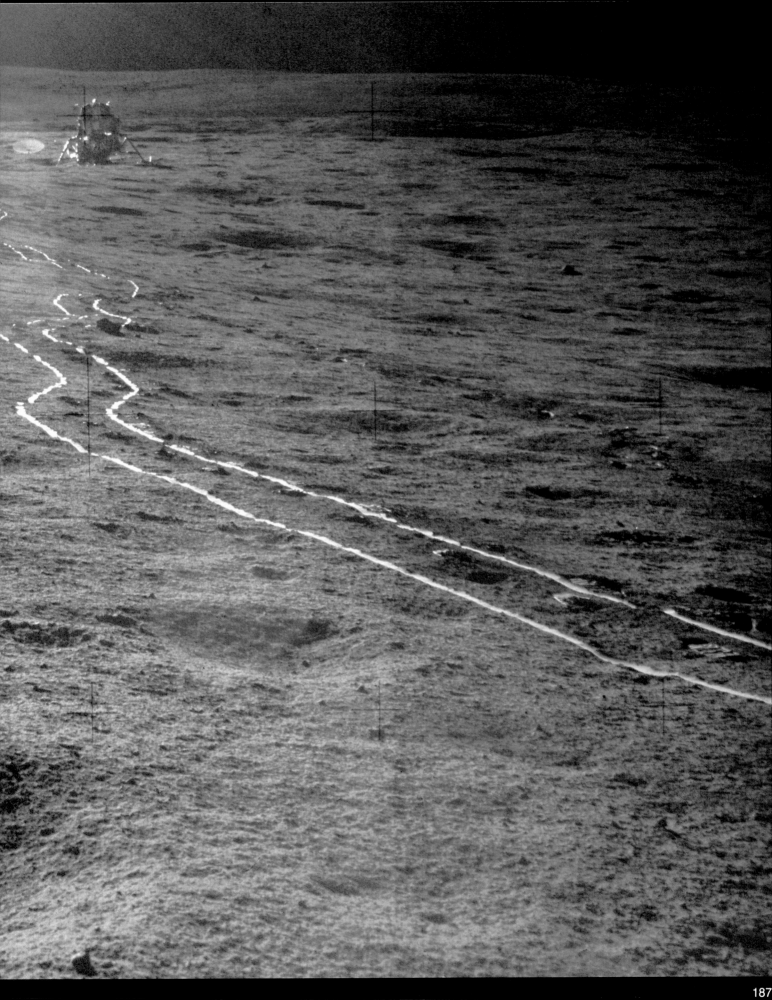

In early August 1971, in the most ambitious moon probe yet, *Apollo 15's* Dave Scott and Jim Irwin roamed the forbidding terrain near a massive 15,000-foot peak named Mt. Hadley, an area cut by treacherous gorges and studded with strange mounds and craters. Luckily they had the use of a remarkable new electric car that looked like an earthly dune buggy—the Lunar Roving Vehicle. The first powered vehicle to be driven on the moon, Rover arrived tucked into the side of the lunar module. "Buckin' bronco!" yelled Irwin as they set off on a series of 8-mph sweeps that would take them through 18 miles of lunar highlands. They ventured to the rim of the Hadley Rille, a 1,200-foot-deep canyon that geologists believe was scoured early in lunar history by fast-flowing molten lava. In their 67-hour stay, the astronauts performed a multitude of experiments. Scott drilled holes that showed the moon's substrata were far firmer than expected, and the astronauts gathered 175 pounds of rock, including a piece of anorthosite that later proved to be more than four billion years old—older than any rocks ever found on earth. It was dubbed the "Genesis Rock."

At the edge of the Hadley Rille, with Mt. Hadley rising beyond, Jim Irwin tends to the ungainly but precious lunar rover. In spite of Rover's slow speed, the astronauts were cautioned to buckle their seat belts: In the moon's gravity, even a slight bump could toss them out.

As one astronaut worked near the lunar module (below, left), the other took a series of pictures to create this sweeping panorama of the barren moonscape. Similar views were shot for TV viewers on earth with a camera mounted on the lunar rover itself and remote-controlled by technicians back in Houston.

APOLLO 16

As a follow-up to its predecessor, *Apollo 16* had the main objective of probing a different region of the moon's surface—the lunar highlands. But the mission yielded more than its share of scientific puzzles. Mission commander John Young and his fellow moon explorer, Charlie Duke, discovered that one area, the Cayley Plains, had five to 10 times the magnetism of surrounding terrain; scientists back home surmised this might be proof that at one time the moon had spun faster than it does now. A rock from the so-called Descartes area turned out to be radioactive, although no one knew precisely why. Finally, the rocks from the Cayley Plains area confounded geologists by being of different composition from what had been expected; further study of them might unlock some of the secrets of the moon's formation. But the mission was not exclusively studious. As their colleague Ken Mattingly in the orbiting command module mapped the far side of the moon, Young and Duke sped around in Rover to set a new lunar wheeled speed record: 11 miles per hour.

Beaming, astronauts Young, Duke and Mattingly acknowledge cheers on the carrier *Ticonderoga* after splashdown. Their good health reassured the physicians; some earlier astronauts had registered abnormal heart rates, but a diet rich in potassium apparently prevented the problem for *Apollo 16's* crew.

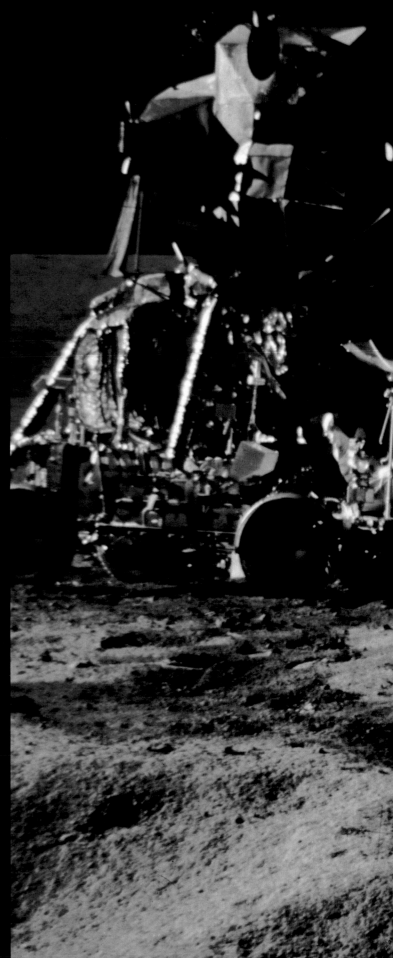

In a burst of exuberance, Young goes aloft for Duke's camera to demonstrate the moon's meager gravity pull. To the left, the lunar rover is parked ready for use beside the lunar module. Young delighted in putting the electric buggy through Le Mans-type skidding turns.

The last and longest lunar landing mission, *Apollo 17* was also the most productive scientifically. During the record 22 hours spent outside on the lunar surface, Gene Cernan and Jack Schmitt drove the lunar rover 22.5 miles around the Taurus-Littrow area of the Sea of Serenity. Schmitt, a geologist, collected samples of basalt rock that were 100 million years older than those found by *Apollo 11*, and he found other rocks that were more than four billion years old—almost as old as the solar system itself. But the most startling find was the orange soil Schmitt discovered at the rim of a crater. Scientists later reasoned that the soil may have been oxidized in a volcanic eruption little more than 200,000 years ago—the most recent lunar geological event they had yet identified. Finally, on December 14, Schmitt and Cernan prepared to lift off for the rendezvous with Ron Evans, orbiting overhead in the command module *America*. "We leave as we came," Cernan said in farewell, "and, God willing, we shall return with peace and hope for mankind."

The launch of *Apollo 17* on December 7, 1972, was the first manned launch to be staged at night. Lift-off was delayed 2 hours and 40 minutes, until 12:33 a.m., because of the failure of an automatic countdown sequence that controlled the last three minutes of the launch.

The last Apollo crew—Jack Schmitt, Ron Evans and Gene Cernan—gather before the launch. Cernan, the only space veteran among them, had been within nine miles of the lunar surface on *Apollo 10's* flight. This time the astronaut would get to land.

Just east of South Massif Mountain, Schmitt inspects a huge boulder that dwarfs the lunar rover. Like all the moon walkers, he and Cernan left a trail of footprints *(lower right)* that will remain for millions of years—uneroded by rain or wind on the weatherless lunar surface.

"We're not trying to bury the patient yet," said one worried NASA official, "but we know it's sick and in the hospital." It was May 14, 1973; the "patient" was Skylab, a 118-foot-long orbiting space station launched by a Saturn V rocket that afternoon.

The first of three teams slated to operate the station was supposed to blast off the next day, rendezvous with Skylab and begin the scientific work that was the justification for this $2.6-billion successor to Apollo. Living for weeks in Skylab's two-story orbital workshop *(page 198)*—the converted third stage of a Saturn V rocket—the astronauts were to survey the earth and study the sun. Above all, they themselves were to be primary objects of study, their every function scrutinized to determine whether human beings really could adapt to life in outer space.

But little more than one minute after lift-off, the space station's aluminum thermal and micrometeoroid shield had broken away, ripping off one of the two winglike solar-cell arrays folded against Skylab's sides. The other wing was pinned shut by a fragment of the shield. When Skylab entered orbit, Houston Mission Control sent signals to open the solar wings. Nothing happened. Twice more signals went out, twice more there was no response. At 90 minutes into the mission, Houston knew beyond a doubt: Skylab was in trouble.

Mission commander Pete Conrad, pilot Paul Weitz and science pilot Joseph Kerwin learned the bad news that night. Their flight had been postponed and might be canceled. Without the wrap-around thermal shield, Skylab's skin was being scorched at 325° F., its innards baked at temperatures up to 165°. Such heat would spoil food, film and medicine—even cause the foam insulation in Skylab's walls to release lethal gases.

To gain some time until a more permanent remedy could be devised, Mission Control fired the small thrusters that ringed the station, to tilt the damaged area away from the sun's direct rays. The average temperature inside Skylab dropped to about 130° F. Meanwhile, engineers worked around the clock designing emergency sunshades. The one chosen was a 22-by-24-foot rectangular parasol with extendible ribs modeled on telescoping fishing rods.

Armed with the parasol and other repair tools, the astronauts finally boarded one of the Apollo spacecraft that would be Skylab's "taxi" fleet and blasted off on the 25th of May, 10 days behind schedule. This first mission had been designed to test whether the astronauts could work in space; now their work would determine whether there would be much of a Skylab program at all.

After a 7½-hour chase Conrad called out, "Tallyho! the Skylab!" and reported that "the area where the shield ripped off looks like it has been scorched black by the sun." After beaming TV pictures to Houston, Conrad docked the Apollo command module nose-to-nose with Skylab.

The "Fix-It Crew" had a quick supper, then Conrad undocked and maneuvered the Apollo alongside Skylab's pinned wing. The astronauts sealed their spacesuits, depressurized the cabin and opened the hatch. Weitz leaned out and, with Kerwin holding his legs, tried to free the wing, using a 10-foot-long pole with a hook attached. The wing did not budge. "We pulled as hard as we could," Conrad said. "Houston, we're going to have to give up. I really feel bad because it's just a tiny half-inch strap. But boy, did it rivet itself into the side of the panel."

After redocking, the crew spent the night in the command module. The next morning, they entered Skylab for the first time, cautiously checking for toxic gases. Mission Control's efforts to minimize the effects of the loss of the heat shield had paid off. No insulation had been damaged, and no fumes were found in the workshop.

By midafternoon the astronauts

Skylab's "solar windmill," projecting from the telescope mount, held 165,000 tiny silicon solar cells that generated 4,000 watts of power—about two thirds the total needed to run a fully operational mission. Of the two side panels, which were to add another 4,000 watts, only one survived the launch.

were set to erect the parasol shield. While Kerwin went into the Apollo to look out the window and help direct things, the other astronauts slipped the shield and its handle out through a small air lock and began adding rods to the handle. They pushed the parasol about 20 feet beyond the skin before extending the four ribs, then pulled the shield back against Skylab's body to help smooth wrinkles that were keeping the shield from opening all the way. At length the parasol was deployed, and Skylab's skin temperature dropped by 50° to 60° F. "We hope that by tomorrow," Mission Control told reporters, "the inside of Skylab will be like Phoenix on a warm, sunny day."

That night the crew moved out of the cramped Apollo command module to sleep in the 17-by-10-foot multiple docking adapter, where the temperature was a cool 68° F. The next day, Sunday, May 27, they unpacked and began to settle in. On Tuesday they activated the eight telescopes in the 12-ton solar observatory mounted on the outside of the space station. That night, with temperatures in the workshop down to 80° F., the astronauts slept in their living quarters for the first time.

The next day they started up the EREP, or Earth Resources Experiment Package—cameras and other instruments that could section and survey a continent while distinguishing a single cornfield or house. The first EREP survey was a success, recording information as Skylab passed over the Southwestern United States, Central America and Brazil. But the survey had taken its toll: Skylab's batteries were severely depleted. The four solar panels atop the telescope mount could not continue to power all of the experiments and Skylab's ordinary functions as well. The pinned solar wing would have to be freed.

Solving the problem required the cooperation of ground crews and engineers. In the Neutral Buoyancy Simulator, where the Skylab astro-

Astronauts Edward Gibson and Gerald Carr peer up from the living quarters of the 48-foot-high orbital workshop. The empty spacesuits standing above them were anchored to the floor of the workshop's upper deck when not in use.

nauts trained underwater to simulate working in zero gravity *(page 202)*, Russell Schweickart and Edward Gibson tested tools and procedures.

On June 7, Conrad and Kerwin donned their pressure suits and went outside. With step-by-step instruction relayed from Schweickart on the ground, they assembled a 25-foot-long pole and attached powerful cable cutters on the end to sever the binding metal strap. Like the Gemini astronauts who first rehearsed making repairs in space, Conrad and Kerwin found it rough work. On the ground, Schweickart compared it to "hanging by your toes" from a trapeze: "There are no normal handholds or footholds." But at length the solar wing was freed, and there was jubilation in Houston as it began to generate power. "Looks like we can turn on more lights," Conrad deadpanned, "and stop living like moles."

Now the astronauts could go about the business of adapting to life in space. They rode an exercise bike 30 minutes a day and took tests to gauge the disorientation of weightlessness—Kerwin even slept with an electronic "bunny cap" to see if his dreams had changed. Floating about the workshop in cleated shoes that could lock into the grid floor, they discovered something at once profound and simple. "You *do* have a sense of up and down," Kerwin reported. "It turns out that you carry with you your own body-oriented world, independent of anything else."

Sleeping and eating were relatively mundane. The astronauts took turns "cooking": On their assigned days they consulted the master menu and laid out such earthly delights as packages of frozen lobster Newburg and filet mignon, which were heated in specially designed trays. They slept in their own closet-sized bedrooms, zipped into sleeping bags stretched against the wall—their heads toward whichever "up" they preferred.

Hygiene was more complicated.

The astronauts used an air-suction toilet, with a lap belt and foot straps to hold them to the seat—but mastering the technique took practice. Weekly showers were taken in an enclosed, collapsible stall, where the water, instead of falling in a stream, broke up into a cloud of floating droplets that splashed on impact. Very little water was needed—only three quarts per shower—but after the shower the water did not just disappear. A vacuum hose was needed to collect the droplets from the curtain and bather—a tedious process.

At length their stay was over. It took two days to clean up, shut down and pack. Then Conrad and his crew climbed aboard and tested the Apollo spacecraft, made a final inspection pass and headed for home.

After spending 28 days and 50 minutes in space, they came back changed men—literally. All three had lost weight, especially in their legs, which had grown thin through a combination of underuse and the redistribution of body fluids. Without the drag of gravity, their spinal columns had stretched, making them each roughly one inch taller. Their hearts had shrunk about 3 per cent and raced when required to pump on earth again. Only their stomach muscles had actually gotten stronger in space, the result of bending without the aid of gravity. The men felt light-headed and heavy-limbed. Kerwin reported an "awful feeling that the world was about to swallow you up."

But two days after splashdown, all had recovered. "If this is the worst that space can do," Kerwin said, "we're up there to stay." NASA head James Fletcher agreed. "Essentially all the objectives of the mission have been completed," Fletcher said. "None of us really dreamed that this could be done at the time the shield failed. The mission has exceeded our wildest expectations."

On July 28—one month and six days after the first crew splashed down—the second Skylab team

blasted off. Their mission was essentially a continuation and expansion of the first one, but it was an ambitious expansion: The projected stay was 59 days, more than twice that of Skylab 1. The three astronauts were Navy Captain Alan Bean—Pete Conrad's pilot on *Apollo 12*—electrical engineer and physicist Owen Garriott and Marine pilot Jack Lousma. Both Garriott and Lousma were rookies, but Lousma was so calm he dozed during the countdown.

Skylab 2 had its own early scare. One of the rocket clusters that controlled the Apollo spacecraft's attitude sprang a leak. "It looks like we've been driving through a snowstorm real fast," Lousma said, watching frozen propellant spray past the window. The astronauts shut down the cluster and the leak stopped. A few days later, when another thruster group sprang a leak, Houston began to plan a rescue mission with backup vehicles that included the craft set aside for the joint U.S.-U.S.S.R. Apollo-Soyuz mission *(pages 222-225)*. NASA even made plans to ask the Soviets for help if all else failed. But the leaks in the two systems proved unrelated and the other two rocket clusters remained healthy.

The most personally debilitating problem hit the astronauts almost as soon as they arrived at the space station. As they began unloading the command module, spacesickness struck. All three felt queasy, and Lousma vomited. For two days the crew was scarcely able to move. No one felt like eating. Motion-sickness pills and head exercises to steady the inner ear were little help.

But by the fourth day the astronauts were on the mend. They began setting up 19 experiments—winners in a nationwide high-school science contest—and found that six mice and some fruit-fly pupae had suffered a fate worse than spacesickness: They had died. Other species fared better. A pair of minnows at first swam in tight loops but soon adapted—they used the wall behind their bag as the "bottom" and swam in line with it. Two spiders—Anita and Arabella—reacted similarly, first spinning disjointed webs but later managing normal ones. But the fastest learners were the minnows hatched in space. They swam normally from the start, "as if they'd adapted while still in the egg," said Garriott.

The astronauts adapted well—once their spacesickness subsided—both to weightlessness and to the work load. On their 10th day, Garriott and Lousma attached a second sunshade to reinforce the aging parasol, and all three crew members spent long days at the solar observatory's display console.

Moreover, they exercised twice as much as the first crew, keeping up strength for their longer stay. Tests every three days showed that they too were losing weight, bone calcium and red blood cells. But after about 40 days a curious thing happened: A plateau seemed to have been reached, and the losses stopped.

In spite of their sickly start, the second Skylab crew completed 39 earth surveys and logged 305 hours at the solar telescopes—50 per cent more than expected in both cases. By the time they shut down the lab and went home at the end of their 59 days,

they had traveled 24.5 million miles. Almost all of their remarkable records were broken, however, by Skylab 3, the last and longest of the missions. Several problems, including cracked stabilizing fins on the Saturn IB launch vehicle, delayed launch for six days. But at 9:01 a.m. on November 16, 1973, the all-rookie crew of Marine Lieutenant Colonel Gerald Carr, engineer and physicist Edward Gibson and Air Force Lieutenant Colonel William Pogue lifted off for their 12-week sojourn in space.

To avoid spacesickness, the astronauts slept in the Apollo craft their first night up. Then they set to work. Soon both the astronauts and Mission Control were complaining about the schedule—astronauts that they had no time to relax, controllers that not enough was getting done.

There were two reasons for the problem. First, the schedule of experiments for this final mission had been increased in the weeks before launch, but no additional training had been given to the flight crew. Second, many of the 20,000 or so pieces of equipment in the workshop had been misplaced by the two previous crews, making experiments difficult to perform efficiently. The astronauts felt overwhelmed by the work load and made a number of errors. After some frank discussion with Mission Control, the work schedule was adjusted—and by the end of their stay, the third crew had surpassed even Mission Control's expectations.

During their time aloft, the astronauts watched the seasons change in two hemispheres, as wheat ripened in the Argentine summer and northern rivers grew swollen with ice. They found nearly fluorescent rivers of plankton off both coasts of South America, photographed a smoking volcano in Japan and at night, said Carr, saw the Eastern United States light up "like a spider web with water droplets on it."

On Christmas Eve, the men found presents from their wives that had

been hidden on board, and they joked that the North Pole looked busy—"so my kids have got some hope," Gibson said, "but they had better keep in line for another day." On Christmas Day, Pogue and Carr took a record seven-hour space walk to photograph the comet Kohoutek.

Using the Skylab smelter, the crew manufactured alloys and crystals stronger and more uniform than any made in earth's deforming gravity. Then, after more than two months aloft, Gibson photographed something that all three Skylab crews had sought: the birth and life of a solar flare. The study of solar flares—violent eruptions on the sun's surface—might help efforts to control nuclear fusion as a source of energy on earth.

Finally, in early February 1974, the astronauts began to close Skylab. They made a last EVA to collect film and cameras and then packed up the Apollo "taxi." "It's been a good home," Gibson said, looking back at the space station. "I hate to think we're the last guys to use it."

The end of the mission aloft was just the beginning for scientists down below, who would be evaluating the data from Skylab for years. The more than 180,000 photographs of the sun, for example, revealed a star much more complex than had been thought. And the earth surveys uncovered new deposits of oil, ore and water. Medical data on the crews suggested a surprising conclusion: Space was not only "kind to people," as Kerwin said on his return, but seemed actually to get kinder with time. The third crew came back even healthier than the second and readapted to earth's gravity faster.

Skylab itself kept going for five and one half years more, its orbit gradually decaying. Then, on July 12, 1979—after a brief period back in the world's headlines—Skylab plummeted through the earth's atmosphere. To everyone's relief it scattered over uninhabited ocean and Australian desert. □

Skylab 1's "Fix-It Crew" blasts off aboard a screwdriver on their way to repair and inhabit the space station in this cartoon published in May of 1973. "The Mercury and Apollo programs gave us our Columbuses," said one NASA official. "Now we need our Jamestown settlers."

LEARNING TO LIVE ALOFT

SKYLAB WORKSHOP
AIR-LOCK MODULE
MULTIPLE DOCKING ADAPTER
APOLLO TELESCOPE MOUNT
PAYLOAD SHROUD

S-II SECOND STAGE

S-IC FIRST STAGE

An artist's sketch shows the 118-foot-long Skylab atop a mammoth Saturn V rocket that dwarfs the human figure shown at its base. The space station, thrust into orbit on the 14th of May, 1973, consisted of *(in cutaway, from the top)* the Apollo telescope mount, the multiple docking adapter, the air-lock module and the orbital workshop where the crew spent most of its time *(also shown in detail above right)*.

Unlike earlier astronauts, the three-man crews assigned to the 1973 Skylab program regarded themselves not as explorers but as pilgrims: the pioneer settlers of space. Once in orbit, they intended to stay for a while—at least four weeks—before returning to earth. Their remarkable, 100-ton spacecraft was by far the largest and most complex yet built. The interior *(right)* of its two-story orbital workshop alone contained more than 10,000 cubic feet and boasted as many of the mundane comforts of home as NASA's engineers and scientists could provide.

What could not be anticipated, however, were the physical and psychological hazards of being weightless for so long. It was feared that, without having to pull against gravity, the astronauts' hearts would grow lazy, their muscles atrophy and their bones lose calcium.

For months preceding Skylab's launch, the crews trained intensively under conditions made as realistic as possible. Using full-scale simulators and mockups, they practiced the routines and chores of daily living in space. They also learned to operate the sophisticated devices that would measure the effects of prolonged weightlessness on their bodies.

Although one member of the first Skylab crew, Joseph Kerwin, was a qualified physician, all crew members received practical instruction in coping with medical problems. In a hospital emergency room they learned to stitch up cuts, and in a dental clinic they practiced pulling teeth—which was deemed the best way to cure a toothache in space.

The condition hardest to simulate was zero gravity. But underwater, in carefully weighted suits *(page 202),* they could come close. The hours of working in neutral buoyancy would prove crucial to solving the problems that lay ahead.

Weightless astronauts perform their daily tasks in this cutaway drawing of Skylab's double-decker orbital workshop, converted from the third stage of a Saturn rocket. The floors and ceilings of the workshop are open grids into which floating astronauts can hook cleats attached to the bottoms of their shoes, in order to stand—or hang—in place.

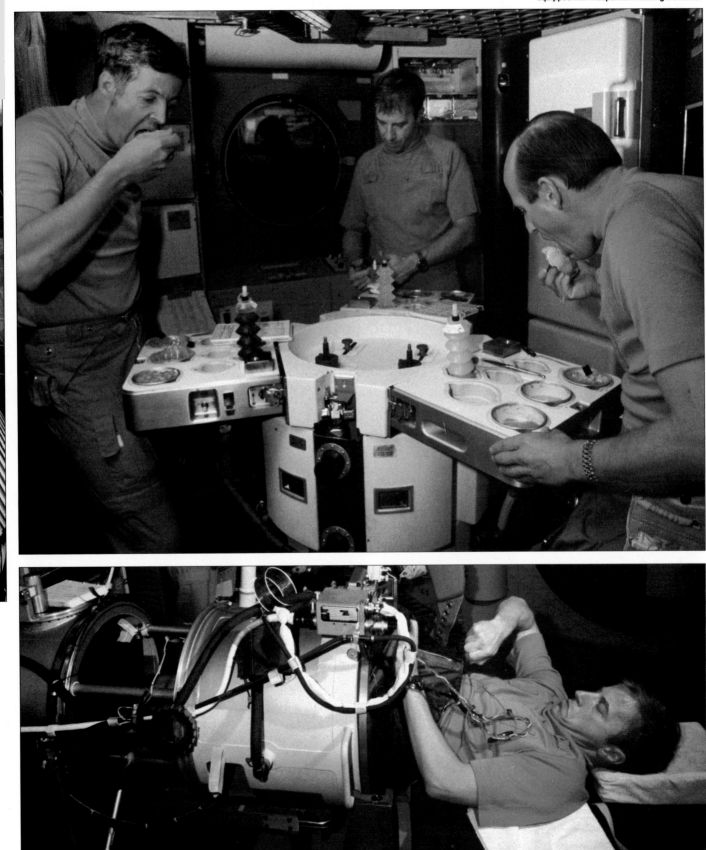

Inside a mockup of their wardroom, the first Skylab crew of *(from left)* Joseph Kerwin, Paul Weitz and Pete Conrad dine from trays equipped with independent heating elements.

Astronaut Weitz seals himself into the Lower Body Negative Pressure Device, a machine that will monitor any change in his circulatory system produced by weightlessness.

Inside a mockup of Skylab's multiple docking adapter, Pete Conrad practices an experiment to determine whether metal can be welded in a weightless environment. The experiment—putting metal samples in a small vacuum furnace—would show if objects made of metal could be assembled and repaired in space.

Conrad mops his brow after a stiff workout on an exercise bicycle. The wheelless machine would be the Skylab crews' principal form of exercise while in space and would also monitor any deterioration in their leg muscles.

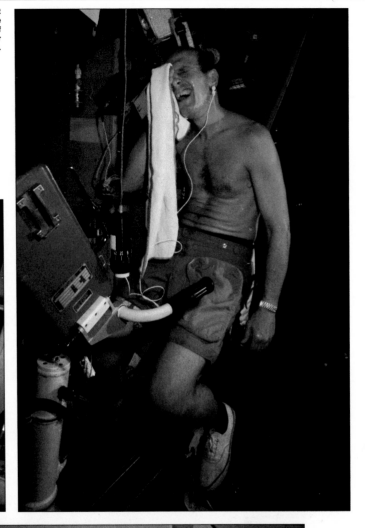

Commander Conrad learns how to operate a motorized chair designed to measure his susceptibility to dizziness and disorientation in zero gravity. The sphere and pointer he holds were used to test his hand-eye coordination.

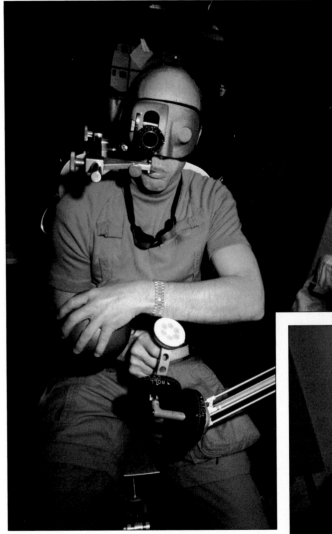

Science pilot Joseph Kerwin tucks himself into the oscillating chair of a special scale devised to determine weight at zero gravity. Using Newton's formula that force equals mass times acceleration, the astronauts were to check their weight every day by measuring how much a known force accelerated their bodies.

The one place on earth where Skylab crews could work for long periods in an environment that simulated weightlessness was underwater. At Marshall Space Flight Center in Huntsville, Alabama, Conrad, Kerwin and Weitz spent hundreds of hours training in a 1.3-million-gallon tank, 75 feet across and 40 feet deep—large enough to accommodate full-sized mockups of Skylab.

The astronauts added lead weights to their pressurized spacesuits until they achieved neutral buoyancy: Weighing as much as the water they displaced, they neither sank nor rose in the water. The buoyancy of their tools—hammer, pliers, C clamps—was also neutralized by the attachment of flotation units.

As a result of their training, the astronauts were at least theoretically prepared to do repair work in space, an ability that took on a new urgency when a mishap just 63 seconds after Skylab 1's lift-off on May 14, 1973, damaged the unmanned workshop and threatened to turn it into a derelict. The astronauts' launch was postponed, and the crew returned instead to the test tank in Huntsville. There, using specially made tools in a race against time, they practiced making the repairs that they hoped would save the mission. □

Pete Conrad, who is dressed for space, goes underwater instead in the Neutral Buoyancy Simulator at Marshall Space Flight Center. Below, with NASA engineers in scuba gear, Conrad and Joe Kerwin practice deploying a solar shield on a Skylab mockup.

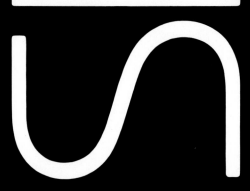

SKYLAB

MISSION PORTFOLIO

The mission was already in grave jeopardy when Skylab's first crew—Pete Conrad, Paul Weitz and Joe Kerwin—suited up on May 25, 1973, to join the unmanned space station, launched 11 days earlier. Skylab's thermal and micrometeoroid shield had torn loose, ripping off one solar-cell array and pinning the other to the side of the craft. Before any of the mission's scientific work could begin, the crippled space station had to be repaired. "We can fix anything!" Conrad boasted—and the astronauts made good on the boast. They installed a parasol to replace the missing shield, and—in a 3½-hour EVA—freed the jammed solar wing so it could generate the power needed to perform the mission's experiments and maintain the space station's life-support systems. After an American record of 28 days aloft—during which their physical condition was carefully monitored—the "Fix-It Crew" returned home, leaving the workshop in orbit, ready for the visit of Skylab 2.

Each carrying a portable air conditioner in a compact case, astronauts *(from left to right)* Joe Kerwin, Paul Weitz and Pete Conrad leave their quarters for the brief ride to the launch pad at Kennedy Space Center on May 25, 1973.

A few seconds after lift-off, a Saturn IB rocket thrusts the Apollo command module carrying the Skylab 1 crew toward a rendezvous—to take place nearly eight hours later—with the space station orbiting 270 miles above the earth.

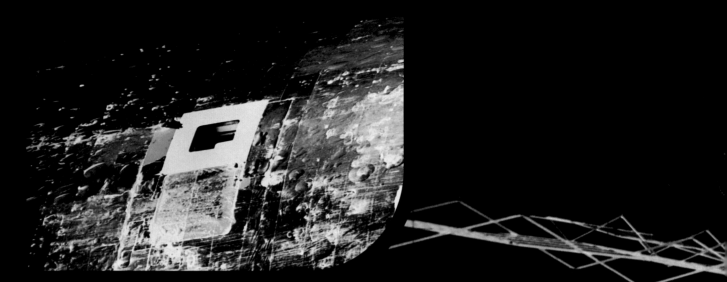

Skylab's unprotected surface is blistered and
tarnished from 11 days of direct exposure to the
sun at temperatures as high as 325° F. The air
lock *(center)* opens on the orbital workshop;
inside, the heat threatened to ruin film and food.

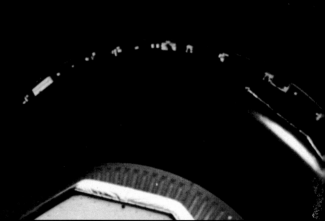

As they neared the damaged space laboratory
before docking, the astronauts were able to see
the telescope mount *(rear)* and its windmill-
like solar panels fully deployed. But, as feared,
the partially open solar wing at bottom right
was jammed, and only broken cables remained
of the missing wing at upper left.

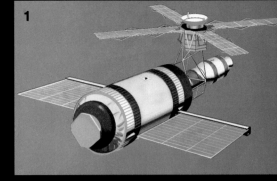

1

2

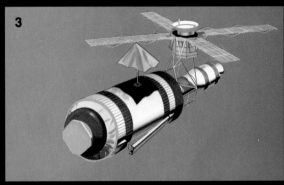

3

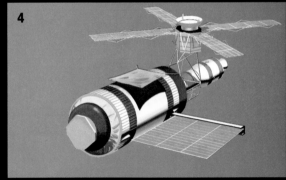

4

At top in this series of drawings, Skylab orbits as intended, with both of its electricity-generating solar wings fully extended. In reality (2), the left-hand wing had been sheared away and the one at right was clamped to the hull. Once inside the sweltering ship, the astronauts extended an enormous parasol with collapsible ribs through a small air lock (3) and then lowered it to shade the hull. Later they were able to free the remaining wing (4).

Repaired, Skylab circles a cloud-covered earth in this stunning photograph taken from the Apollo command module near the end of the first crew's 28 days in space. The surviving solar wing is extended and the improvised parasol shades the workshop from the sun's fierce rays.

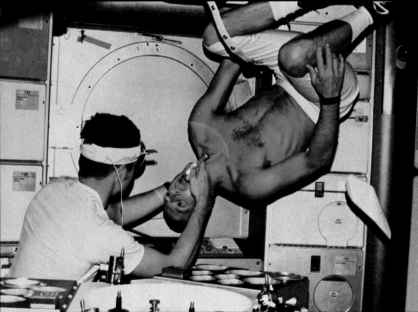

Skylab's collapsible shower stall as a washcloth floats above his right hand. Drying oneself and the stall with a hand-held vacuum hose made showering a tedious, hour-long process.

His feet floating overhead, Conrad pedals the bicycle ergometer with his hands. Exercising hard in the hot workshop at times caused Conrad's heart to skip a beat. Nevertheless, he and the other Skylab 1 astronauts reported that they were not getting sufficient exercise.

Suspended from the ceiling by a single knee strap, Conrad undergoes an oral examination by Dr. Kerwin. "You do have a sense of up and down," Kerwin reported. "You say to your brain, 'I want that way to be up,' and your brain says, 'Okay, then that way is up.' "

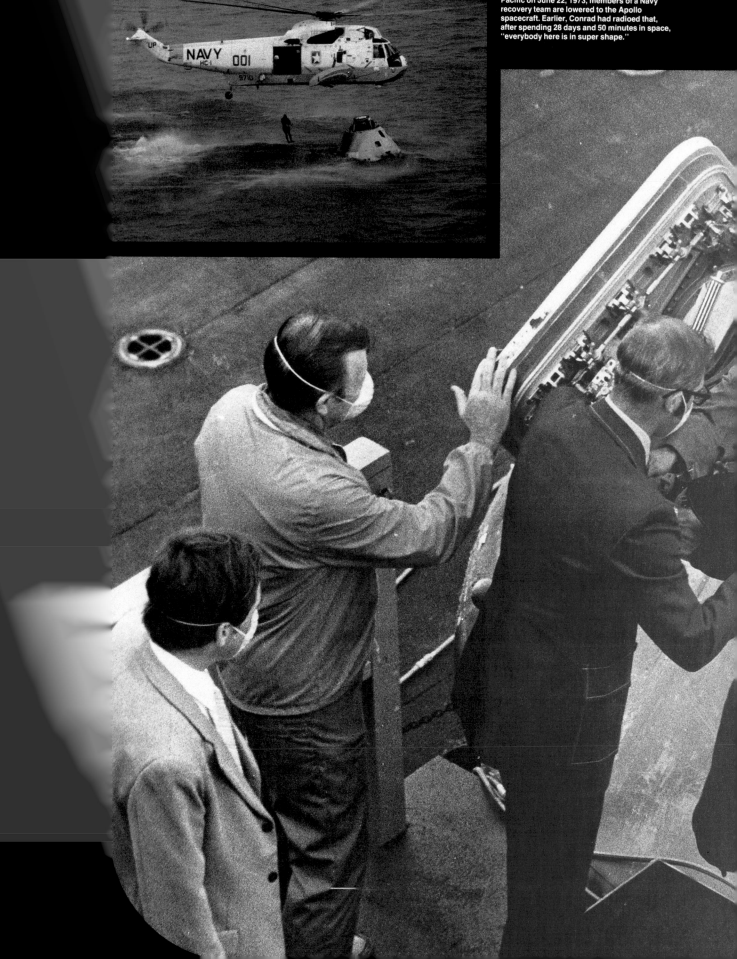

ew minutes after splashdown in the South Pacific on June 22, 1973, members of a Navy recovery team are lowered to the Apollo spacecraft. Earlier, Conrad had radioed that, after spending 28 days and 50 minutes in space, "everybody here is in super shape."

Conrad is the first one to emerge from the hatch of the command module. Members of the NASA welcoming party wear face masks to shield the astronauts from earthly germs that might confuse the study of their physical condition.

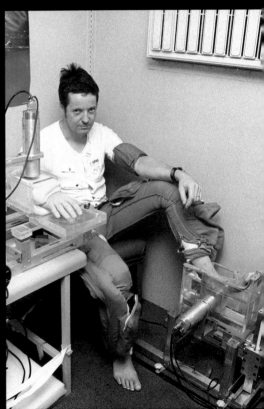

A pale, tired-looking Joe Kerwin undergoes tests aboard the *Ticonderoga* to determine the extent of his bone-mineral loss. Although it took several days for Kerwin to recover his strength, doctors reported the astronauts were "in better shape than we had hoped for."

Simply put, the crew of Skylab 2 set out to surpass every performance standard established by their predecessors in Skylab 1. The orbital workshop had been vacant for 36 days when Alan Bean, Owen Garriott and Jack Lousma arrived on July 28. At first they had motion sickness: "We're just not as spry up here right now as we'd like to be," said mission commander Bean. But as the symptoms wore off, the crew became workaholics, putting in 12-to-16-hour days studying the sun and earth, conducting scores of experiments—and asking Mission Control for more. When they returned after 59 days, they had voyaged 24.5 million miles. And they had taken an astounding number of photographs—more than 70,000—of the sun in all its violent moods.

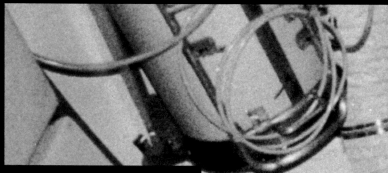

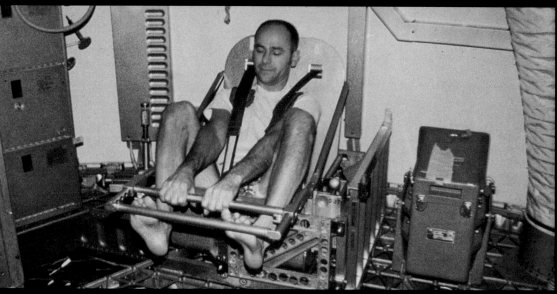

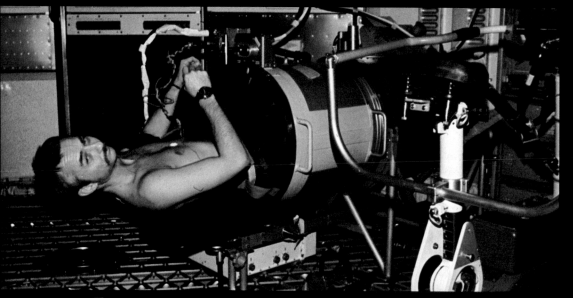

Navy Captain Bean *(top)* weighs himself on Skylab's special scale. Despite eating relatively well, Bean lost 8½ pounds on the trip. At bottom, science pilot Garriott uses a device to monitor his heart's reduced capacity to pump blood. The Skylab 2 astronauts found that changes in their bodies' functions leveled off after 40 days at zero gravity.

Arabella, one of the first two spiders in space, successfully spins a web in weightless conditions, an experiment suggested by a high-school girl from Massachusetts. The crew conducted 19 such student experiments.

Taking advantage of the orbital workshop's roominess, Jack Lousma tries out a jet-powered flying machine, one of three maneuvering units the Skylab 2 astronauts tested for possible future use in making repairs in space.

Seen from above, a second sunshade *(top)* spreads over the first, secured to the telescope mount *(center)* by two metal poles. Once it was in place, temperatures in the orbital workshop dropped to the comfortably low 70s.

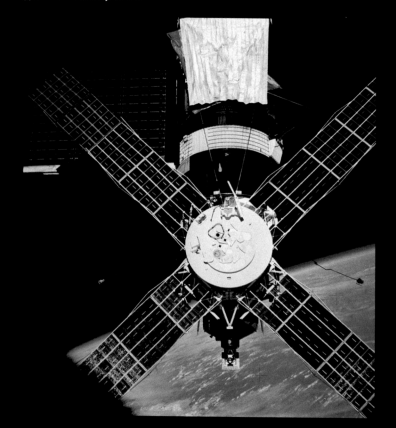

A portion of Skylab and a gray-blue earth are reflected in Jack Lousma's visor during the record 6½-hour-long space walk in which he and Owen Garriott unfurled a second sunshade over the spacecraft's "bald spot."

Garriott clings to the Apollo telescope mount just after setting up a device for collecting micrometeoroids during his first EVA. On this space walk, the 42-year-old Oklahoman

When they ascended smoothly from Cape Canaveral in mid-November 1973, the trio of rookie astronauts manning Skylab 3 knew they would not be home for the holidays. But, building on the experience of those who went before them, Gerald Carr, William Pogue and Ed Gibson were determined to avoid some of the hazards that afflicted their predecessors—and to find time to enjoy themselves during a grueling 84-day voyage. They took medicine to ward off motion sickness, packed an extra 160 pounds of groceries, and carried a compact treadmill to add variety to a stepped-up exercise program of 90 minutes per day—three times the first crew's regimen.

During their time aloft the crew achieved some significant firsts. They made high-quality alloys and crystals in their electric furnace, they took turns going outside to change film in the telescope-mount camera and to photograph the comet Kohoutek on its loop around the sun, and they filmed the spectacular birth of a solar flare *(opposite)*.

The birth of a solar flare—an awesome eruption of energy spanning one third of a million miles along the sun's fiery rim *(upper left)* —is captured on film with extraordinary clarity from above earth's obscuring atmosphere by Skylab 3's science pilot, Ed Gibson.

Brilliantly illuminated by arc lights, the Apollo spacecraft for Skylab 3 and its Saturn IB launch vehicle are reflected in the tranquil waters off Cape Canaveral in the hours before lift-off on the 16th of November, 1973. At left is the mobile service structure.

Wrapped in a rubberized flotation collar, the
Skylab 3 command module with the astronauts
still aboard is hoisted to an auxiliary deck of
the recovery ship U.S.S. *New Orleans* after
splashdown in the Pacific on February 8, 1974.

On board the *New Orleans,* astronauts *(from
left)* Edward Gibson, William Pogue and Gerald
Carr are attended by a masked medical team.
After 84 days of weightlessness, the trio at first
found every physical effort a chore. "I felt,"
said Pogue, "as if I weighed a thousand tons."

A bare-chested Pogue tests his endurance on
an ergometer bike as part of the crew's postflight
examination. Probably because they exercised
more, Skylab 3 astronauts did not suffer
the loss of muscular strength or weakening
of the heart experienced by earlier crews.

During a last fly-around in their Apollo
command module before they headed home,
the Skylab 3 crew took this picture of the space
station against a black sky. An American flag
is visible on the docking port they have just left.

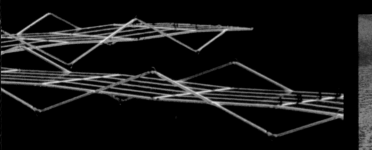

"It's been a good home," said Ed Gibson of the vacated Skylab. Missing a solar-cell wing but with its shades still in place, the space station orbits a cloud-shrouded earth—as it would continue to do for the next five years.

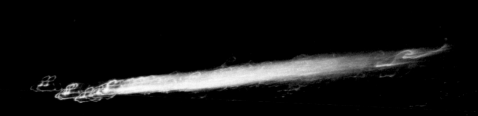

Skylab appears as a blazing streak in this time exposure taken by an Australian sheepman as the craft finally reentered earth's atmosphere and disintegrated on July 12, 1979. Its debris was scattered across hundreds of miles of the Indian Ocean and the Australian Outback.

JOINT MISSION

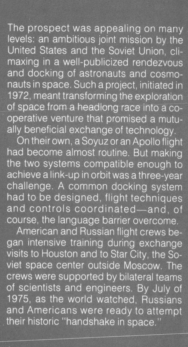

The prospect was appealing on many levels: an ambitious joint mission by the United States and the Soviet Union, climaxing in a well-publicized rendezvous and docking of astronauts and cosmonauts in space. Such a project, initiated in 1972, meant transforming the exploration of space from a headlong race into a cooperative venture that promised a mutually beneficial exchange of technology.

On their own, a Soyuz or an Apollo flight had become almost routine. But making the two systems compatible enough to achieve a link-up in orbit was a three-year challenge. A common docking system had to be designed, flight techniques and controls coordinated—and, of course, the language barrier overcome.

American and Russian flight crews began intensive training during exchange visits to Houston and to Star City, the Soviet space center outside Moscow. The crews were supported by bilateral teams of scientists and engineers. By July of 1975, as the world watched, Russians and Americans were ready to attempt their historic "handshake in space."

Mission commanders Tom Stafford *(left)* and Alexei Leonov stand behind their crews— *(seated from left)* Deke Slayton, Vance Brand and Valeriy Kubasov—during the Russians' visit to Houston in February 1975. The mission would be the first space flight for Slayton, a Mercury astronaut who had been grounded at one time by an erratic heart rate.

The Apollo spacecraft that is carrying Tom Stafford, Deke Slayton and Vance Brand thunders upward from the launch pad on the 15th of July, 1975, approximately seven hours after the launch of Soyuz *(right)*.

The Apollo *(left)* and the Soyuz *(right)* maneuver toward link-up in this simulation made from photos that each crew took of the other craft. From each ship extend three finger-like petals, part of the U.S.-made docking system.

In the first launch ever televised live by the Russians, the two-man Soyuz spacecraft lifts off from Baykonur Cosmodrome in Kazakhstan. The launch pad was the same one the Soviets had been using since Sputnik in 1957.

Stafford *(left)* and Leonov, carrying a camera, meet in the padded hatchway of the docking module that connects their spacecraft. During the 44 hours the craft remained linked, the crews exchanged visits and gifts and shared a meal of borscht (from tubes) aboard Soyuz.

On a triumphal tour of the United States, the returned Apollo astronauts and Soyuz cosmonauts are escorted through the streets of Chicago by Mayor Richard Daley. In the lead car, Stafford and Leonov flank Soviet space official General Vladimir Shatalov.

"We have succeeded!" exulted Tom Stafford in awkward Russian when the Apollo, with a gentle bump, locked onto Soyuz as illustrated here by two back-up craft on display in Washington, D.C. His counterpart, cosmonaut Alexei Leonov, responded in English, "Good show!"

In the spring of 1981, the space shuttle *Columbia* was poised for its maiden flight. The launch was more than two years behind schedule, and nearly six years had passed since an American had last ventured into space. Now the future of a hoped-for "second space age" rested on *Columbia*. Spacefaring would either become as dependable, productive and routine as aviation itself or wither into an expensive memory.

An unlikely-looking machine to bear this burden, *Columbia* was a hybrid of evolving technology. The first of an anticipated fleet of at least four such craft, the shuttle was a true aerospace vehicle. The delta-winged orbiter, roughly the size of a DC-9, was secured to an even larger detachable fuel tank; bolted to the tank were two solid-propellant rocket boosters, also detachable. Designed to take off like a rocket and maneuver like a spaceship, the orbiter would glide to a dead-stick touchdown on land as matter-of-factly as an airplane *(pages 232-233)*. The hosts of recovery ships and aircraft that had greeted every American space mission from Mercury through Apollo-Soyuz would never be needed again.

An even more revolutionary aspect of the shuttle was that, unlike all previous spacecraft, it was intended to be reused. After a flight, it would need only modest refurbishing to get it ready for the next mission.

The idea of reusable craft was not new. Wernher von Braun, as practical as he was visionary, had proposed reusable spaceships in his speculations about moon bases and round trips to Mars. Moreover, as early as the mid-1950s, the U.S. Air Force had been experimenting with a series of aircraft built to fly at supersonic speeds at very high altitudes. One of the later craft in this series was the X-15. Carried aloft under the wing of a B-52 bomber, the X-15 could rocket 50 miles above the earth—to the lower reaches of space but not quite into orbit—and then land at Edwards Air Force Base in California.

Columbia—or something like it—had clearly been a working concept for years. But the launch of *Sputnik I* in 1957 had changed the course of the American space program. Rather than take the time to further the development of manned glide rockets such as the X-15, NASA had opted for the "quick and dirty" approach to putting a man into orbit: the ballistic space capsule that was Mercury, which then evolved into the Gemini and Apollo spacecraft.

By the early 1970s, however, reusability was essential if the space program was to grow from single-shot spectaculars to economic feasibility. NASA began developing the shuttle to be the first cargo ship in space. With its roomy flight deck and cavernous cargo bay *(pages 230-231)*, the shuttle could haul people as well as a variety of payloads into orbit. It could deliver communications, weather and military satellites more efficiently than they could be launched by single-use rockets from earth—then return to service and repair the satellites.

Another early assignment projected for the shuttle was to carry aloft a fully equipped scientific laboratory. Called Spacelab, it was developed by the European Space Agency (ESA), whose member nations would share the costs and benefits. Spacelab, like the shuttle, would be reusable; after each flight, it would be removed and reequipped.

Because the shuttle could carry small payloads as well as large ones, NASA proposed to rent spaces in the commodious cargo bay. For $3,000 to $10,000, universities, corporations, U.S. government agencies, individuals and foreign countries could send approved experiments and inventions into orbit. These "getaway specials," self-contained in aluminum canisters, would simply be turned on and off in orbit by the crew.

The nature of that crew was itself a significant breakthrough for the

space program. American astronauts had always been a breed apart, selected for their ability to endure the dangers of experimental space flight and the hardships imposed by the cramped confines of early spacecraft.

With the shuttle, those requirements changed. *Columbia* was built with working and living space for up to seven persons. Its cabin was pressurized to 14.7 psi—normal air pressure at sea level. Even at launch, no one aboard would be subjected to more than three Gs—less than half the acceleration experienced by the crew of *Apollo 11* on the way to the moon. This was possible because the engines would be throttled back to ease the stress on passengers and on delicate scientific payloads.

Moreover, except for the commander and pilot, those aboard the shuttle would not have to fly the ship, which in any case was to be controlled largely by computers. For the shuttle, engineers had developed a sophisticated complex of five onboard computers—four operating simultaneously, checking one another's performance, and a fifth in reserve. An astronaut could take command, but instructions to the craft had to go through the computers.

In 1976, NASA began recruiting a generation of astronauts that reflected this new reality and new opportunity. Anyone—geologist, physicist, engineer—in reasonably good health, who could contribute to the shuttle's varied missions, might apply to be a mission specialist. These astronauts would not have to perform pilot duties, but they would be responsible for coordinating payload operations and carrying out scientific objectives. Among the 35 selected in 1978 under the new criteria were six women; of the men, three were black and one was a Japanese-American.

A third type of specialist, not rated as an astronaut, was also scheduled to fly aboard the shuttle. These payload specialists would be scientists or technicians given the opportunity to do research in space. Unlike astronaut candidates, who had to undergo a rigorous training and evaluation program for space flight, payload specialists would train intensively for their specific research project but would need only a few weeks' familiarization with zero gravity and emergency procedures in space.

Before anyone could go anywhere, however, the necessary hardware had to be created. For engineers at Rocketdyne—a subsidiary of Rockwell International in California, the prime contractor—the greatest challenge was developing reusable rocket engines, compact but immensely powerful, three for each shuttle craft. Rocketdyne's eventual product was the most efficient liquid-fuel rocket engine yet built. It stood 13.9 feet high and weighed only 6,618 pounds, yet it was capable of 375,000 pounds of thrust at launch. But the new machine was subject during development to more delays and failures than successes. During one 1978 test firing, an untried valve failed and the engine exploded.

Engine development slowed the program by about a year, but the problem that was perhaps more embarrassing to the program involved the shuttle's heat-resistant skin. Unlike returning Apollo command modules, whose protective coating charred and flaked away on reentry, *Columbia* needed a permanent, reusable—but lightweight—heat shield. The solution was to cover 70 per cent of the craft's aluminum surface with lightweight tiles—more than 32,000 of them. Each custom-made silica-ceramic tile was covered with a hard coating of borosilicate glass and glued to a felt pad that was bonded to the spaceship's skin.

Weaker and more brittle than expected, the thermal tiles became a problem. At first they were not properly bonded to the spacecraft, and when *Columbia* was ferried from California to Florida in 1979 on a much-publicized piggyback ride

Bathed in light, the winged orbiter *Columbia* and its external fuel tank and pair of solid rocket boosters are primed for launch at Kennedy Space Center. All except the external fuel tank are designed to be recovered and used again.

atop a Boeing 747, a number of tiles fell off. Others were damaged and had to be replaced. For the next 18 months, *Columbia*—derided by some as "the flying brickyard"—sat in a hangar at Cape Canaveral while workers pulled off tiles, tested them, put on new ones and retested them.

Concern over the reliability of the engines and the life-protecting tiles lent extra excitement to the mood at the Cape as the shuttle underwent final preparations for launch in early 1981. There were persistent hitches—and a tragedy. At one countdown rehearsal, five technicians mistakenly entered a compartment of *Columbia* that was still filled with the pure nitrogen used to replace oxygen as a precaution against fire or explosion at launch; two men died.

In April, a scheduled launch was scrubbed with minutes to go when one on-board computer fell out of sync with the other four—by $1/25$ of a second. Two days later this problem had been resolved, and early on April 12, hundreds of thousands of people were gathered along Cape Canaveral's beaches and roadways to witness the maiden launch of the world's first Space Transportation System (STS). By chance, it was 20 years to the day since the Soviets had lofted Yuri Gagarin into space.

The launch of STS-1 was the first time an American spacecraft was going up without prior testing in unmanned flight. Mission objectives were simple: a safe ascent and return for *Columbia* and crew. For this shakedown flight, the two men in the orbiter's cabin were astronauts of the old school. Flight commander John Young, once a Navy test pilot, had joined the program in 1962. Now 50, he had first flown in space in *Gemini 3,* and in 1972 he had become the ninth man to set foot on the moon. This was his fifth mission—an American record. Pilot Robert Crippen was also a veteran of high-performance military aircraft, but he was a space

rookie. At 43, he had been waiting a dozen years for this moment.

At 7 a.m., *Columbia* lifted off in a blast of flame, smoke and thunder, clearing the 348-foot-high launch tower in six seconds. Two minutes into the flight the solid rocket boosters, their highly explosive aluminum powder spent, burst free with a bang and parachuted toward the Atlantic, where seagoing tugs would retrieve them. A few minutes later, when the craft was still short of orbit, the big external tank was jettisoned. *Columbia* was flying upside down, on the tank's underside, the better to stay out of its way as it spiraled upward and then down to disintegrate over the Indian Ocean.

Commander Young fired *Columbia's* two OMS (orbital maneuvering system) engines *(pages 230-231)* twice to put the astronauts into a circular orbit 152 miles above the earth. The ensuing 55 hours were relatively serene as the shuttle performed the way all had hoped: routinely. While the craft circled the earth 36 times, Young and Crippen enjoyed weightlessness in short-sleeved comfort. They tested the big doors of the empty cargo bay (which remained open, except during testing, to dissipate heat) and put the ship through its computerized paces.

The show-stopper of their mission was the landing: It was a beauty. About 25 minutes after firing the OMS engines to slow down slightly, *Columbia* hit the atmosphere with its nose pitched up at a 40-degree angle. As the craft fell—belly first—at Mach 24.5, heat on the tiled surfaces built to 2,700° F. For 16 minutes radio communication was blocked out. Young and Crippen were riding inside a meteor, but except for noticing a pink glow outside the cabin window, they sensed nothing unusual.

At just under Mach 5, Young took over the controls to fly the last two swooping turns manually for the one-chance-only final approach to Edwards AFB. As *Columbia* slowed to the speed of sound, it set off a double

sonic boom. The ship, its engines no longer functioning, glided steeply toward the sun-hardened desert floor. With Crippen reading out the airspeeds, Young pulled out of the dive and a few seconds later lowered *Columbia's* wheels. "We were targeted to touch down at 185 knots, and the very moment I called out 185, I felt us touch down," Crippen said afterward. "John really greased it in."

Seven months later, in November 1981, *Columbia* was ready to go again. In the cabin were a pair of former military pilots: Joseph Engle, 49, and Richard Truly, 44. Both had been training since the 1960s for their hour in space. Flight commander Engle, in fact, had flown the X-15 as an Air Force test pilot.

There had been delays, perhaps predictable in making ready a used spacecraft. A fuel spill in September had required that almost 400 of the orbiter's heat-shielding tiles be detached, cleaned and reglued. Then, 31 seconds before a scheduled launch in early November, engineers discovered that lubricating oil in the auxiliary power units—the same oil used on the first mission—had clogged two vital filters with gunk.

But on November 12, when lift-off finally occurred, everything went smoothly. A new water-deluge system on the launch pad substantially dampened the shock wave from the solid rocket boosters, reducing damage to the pad; and this time not one tile was lost as *Columbia* ascended.

Trouble, however, lay just ahead. In the first hours of the planned five-day mission, one of three fuel cells powering the craft malfunctioned. Mission safety rules required that the mission be cut back. It fell to Sally Ride, the capsule communicator, to radio *Columbia:* "Bad news. You'll be coming home tomorrow."

Engle and Truly used the remaining time to carry out most of their planned goals, including the first test of a huge mechanical arm, built in Canada and formally designated the

Remote Manipulator System *(pages 230-231).* Operated from the cabin by Truly, the RMS would eventually be used to place satellites in space and snatch them back. Coming home, Engle put *Columbia* through a rigorous series of rolls and other maneuvers, then glided through a 25-knot cross wind to a perfect landing.

Columbia would need a third and a fourth test in 1982 before it could be declared operational. The first of these was launched March 22. During their week in space, commander Jack Lousma and pilot Gordon Fullerton became adept with the mechanical arm. They also accomplished a major objective of STS-3, maneuvering the shuttle to expose parts of it to the sun for up to 80 hours at a time to test its thermal shield.

Though things went well aloft, the weather back home was anything but pleasant. Rain had turned the dry lake at Edwards to mud; the alternate landing site, at White Sands, New Mexico, was obliterated by a sandstorm whipped by 55-mph winds. *Columbia* had enough provisions for three more days. The astronauts used one day waiting for the storm to clear. Then Lousma, a veteran of Skylab nearly a decade earlier, brought his ship "high and hot"—fast and steep. Landing at a speed of 250 mph, *Columbia* rolled across three miles of desert before coming to a stop.

Three months later, precisely on schedule on June 27, *Columbia* roared aloft for the fourth time. On board with commander Ken Mattingly and pilot Henry Hartsfield were

packages from the shuttle's first paying customers: an engineering test for separating biological materials, jointly sponsored by an aerospace firm and a major drug company; and the first of the bargain-rate "getaway specials," a compact collection of nine experiments originated by students at Utah State University.

The U.S. Defense Department was also represented on this fourth flight. Its payload included a cosmic-ray detector, ultraviolet and infrared sensors for gauging the tracks of enemy missiles, and a space sextant designed to enable satellites to navigate without guidance from earth.

The fourth mission's only disappointment was the loss of the $32-million casings for the solid rocket boosters. Evidently their parachutes failed, and the reusable casings disappeared in the Atlantic. Nevertheless, before a cheering crowd of 500,000 on the Fourth of July, Mattingly brought *Columbia* to its first hard-surface landing, on a 15,000-foot concrete runway at Edwards.

NASA now declared the Space Transportation System officially open for business. A second orbiter, *Challenger,* was on hand at Edwards when *Columbia's* fourth flight ended, and two more, *Discovery* and *Atlantis,* were in the works.

In November, *Columbia* flew for a fifth time. In addition to its commander, Vance Brand, and pilot Robert Overmyer, it carried the first of the new class of astronauts: mission specialists William Lenoir, a former electrical engineering professor, and physicist Joseph Allen. During the

five-day mission, Lenoir and Allen helped launch a pair of commercial communications satellites (one American and one Canadian).

Then, in April of 1983, *Challenger* took its place on the flight line while *Columbia* was refurbished. The new ship was lighter, slightly more powerful and able to carry 4,000 pounds more cargo than *Columbia.*

Engine troubles had delayed the launch for two months, but once in orbit, *Challenger* demonstrated anew the U.S. space program's knack for problem solving. With commander Paul Weitz and pilot Karol Bobko at the controls, Story Musgrave and Donald Peterson toured the open bay for four hours—the first U.S. space walk in nine years—testing new spacesuits. But *Challenger's* primary mission—to deploy an expensive package of electronic gear known as the TDRS (tracking and data-relay satellite)—nearly met with disaster. After ejection, the satellite's booster rocket misfired, and for a time the TDRS tumbled out of control and was feared lost.

In the anxiety-filled hours that followed, a Houston ground team was able gradually to regain control. Later, a team at Goddard Space Flight Center began the tricky process of pushing the TDRS into a proper circular orbit by firing its small thrusters in a complex series of measured steps. The effort to save the satellite was a pointed reminder that spacefaring—though dependent on sophisticated computers—still required the human ability to react to emergencies.

Meanwhile, NASA readied *Challenger* for its second flight—which would carry the first American woman, mission specialist Sally Ride, into space as part of a crew of five—and NASA coordinated payload reservations for fourscore additional flights through the 1980s. Looking even further ahead, NASA Administrator James Beggs invited firms to submit proposals for building a permanent space station—to be serviced, of course, by the shuttle. □

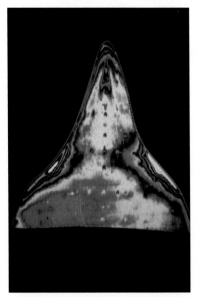

A steel model of *Columbia (below)* is tested in a high-speed wind tunnel to determine areas of greatest heat build-up during reentry. The green and yellow in the computer-enhanced photograph indicate the hottest spots.

Columbia goes in for repairs after a number of its heat-resisting tiles tore loose while it was being ferried from California to Florida in 1979. Each tile is custom-made for a particular spot and requires hours to fit and attach.

Physicist Sally Ride *(above),* 32, was assigned to be the first American woman in space—aboard a shuttle scheduled for June 1983. Fellow astronauts deemed Ride "a very cool operator."

The 122-foot-long orbiter is the one part of the space shuttle that actually shuttles to and from space. The solid rocket boosters and external tank fall to earth after launch; only the orbiter goes into orbit. And of all manned orbiting spacecraft, only the orbiter can fly and land like an airplane.

The orbiter's cabin has an instrument-packed flight deck above cramped living quarters. Two of the flight deck's 10 windows look back over the cargo bay, a 15-by-60-foot hold that can carry as much as 65,000 pounds. The cargo-bay doors hold radiator panels that shed heat built up in flight by vehicle and crew. Cradled inside one door is the Remote Manipulator System (RMS), a 50-foot-long jointed arm that snares payloads with a rotating-wire device called an end effector. The RMS operator shifts and deploys the payloads from the aft crew station.

The orbiter's three main engines form a triangle in the rear. They feed off the large external propellant tank through a network of valves, pipes and pumps, generate 375,000 pounds of thrust each at launch and can fly 55 missions without a major overhaul. Flanking them are the two 6,000-pound-thrust engines of the orbital maneuvering system (OMS), used for major maneuvers such as the transfer from one orbit to another. Both the OMS engines and the 44 small engines of the reaction control system (RCS) use propellants (fuel and oxidizer) that are forced into the engines by helium gas. There are 38 primary RCS engines, each providing 875 pounds of thrust, and six 25-pound-thrust vernier engines. The RCS thrusters are used for small rotational maneuvers (pitch, roll and yaw).

The stubby wings span 78 feet from tip to tip. Each has two elevons—combined elevators and ailerons. A flap below the tail protects the main engines from reentry heat and helps control pitch. The tail holds the vertical stabilizer and the rudder, which fans out to help brake the orbiter at landing.

END EFFECTOR

HELIUM TANK

OMS FUEL TANK

OMS OXIDIZER TANK

RMS ARMREST

RCS FUEL TANK

RCS OXIDIZER TANK

RUDDER

VERTICAL STABILIZER

OMS ENGINES (2)

RCS PRIMARY ENGINES (12)

BODY FLAP

RCS VERNIER ENGINES (2)

MAIN ENGINES (3)

ELEVONS

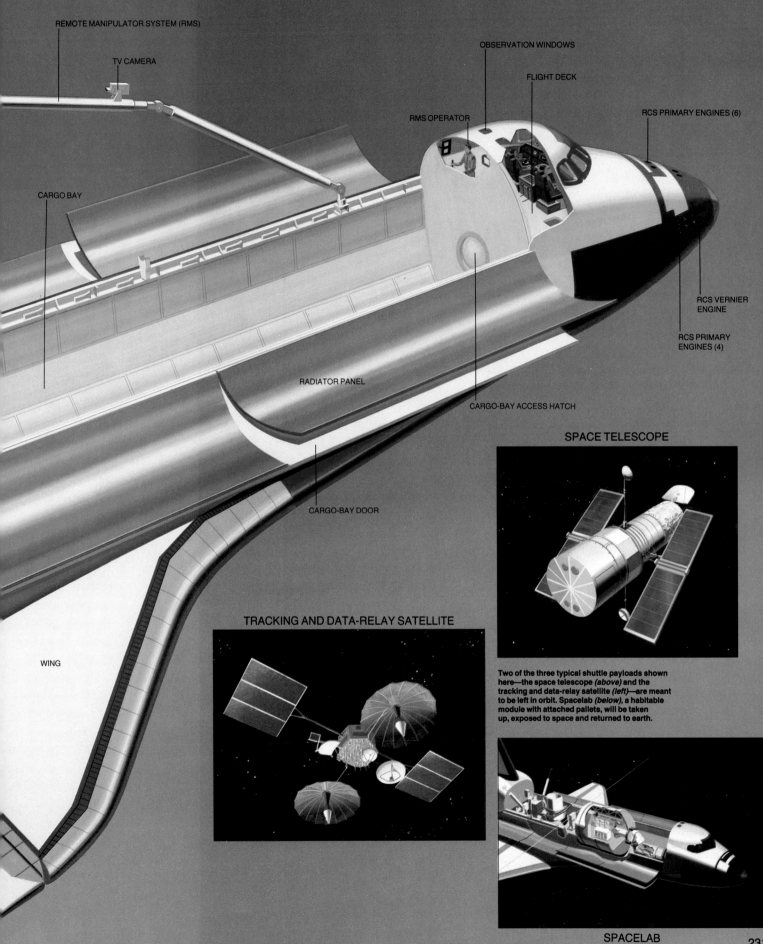

REMOTE MANIPULATOR SYSTEM (RMS)

TV CAMERA

OBSERVATION WINDOWS

FLIGHT DECK

RMS OPERATOR

RCS PRIMARY ENGINES (6)

CARGO BAY

RCS VERNIER ENGINE

RCS PRIMARY ENGINES (4)

RADIATOR PANEL

CARGO-BAY ACCESS HATCH

SPACE TELESCOPE

CARGO-BAY DOOR

TRACKING AND DATA-RELAY SATELLITE

WING

Two of the three typical shuttle payloads shown here—the space telescope (above) and the tracking and data-relay satellite (left)—are meant to be left in orbit. Spacelab (below), a habitable module with attached pallets, will be taken up, exposed to space and returned to earth.

SPACELAB

A firing of the two OMS engines (7) puts the orbiter into a low elliptical orbit. A second burn lifts it into a circular orbit. The cargo-bay doors are opened so that heat from the on-board equipment can escape. If the payload includes earth-sensing devices, the orbiter flies with the cargo bay facing earth (8).

FLIGHT OF THE SHUTTLE

The three main engines fire rapidly in succession at 120-millisecond intervals. Then the solid rocket boosters ignite and the shuttle lifts off, rolls 120 degrees to its right and arcs over the Atlantic Ocean on its back. At T-plus-2 minutes the empty boosters fall away, followed about six minutes later by the external tank. The smaller OMS engines fire to put the shuttle into an elliptical orbit, then fire again half an orbit later to establish a circular orbit about 250 miles above the earth.

Systems are checked, the cargo-bay doors opened and the prescribed number of orbits flown. To prepare for reentry, the doors are closed. RCS jets yaw the orbiter about, and the OMS engines in the tail fire to brake the craft, beginning the hour-long descent. Another burst from RCS jets pitches the orbiter over, nose up, so that the black-tiled belly can protect the craft from reentry heat.

Gradually the spaceship dies and is reborn a glider. RCS jets give way to aeronautical surfaces—first elevons, then rudder—as air pressure builds up. Broad, banking S turns control the rate of descent. The orbiter sinks at an angle seven times as steep as that of a jet-liner—and gets only one try at landing. The nose pulls up just before 1,700 feet; the landing gear drops at 90 feet and—rear wheels first—the shuttle touches down 14 seconds later. □

Two minutes after lift-off (1), explosive bolts and small rocket motors jettison the solid rocket boosters (2), which then parachute the 30 miles to the ocean (3) for recovery and reuse. The huge external tank continues to fuel the main engines for another six minutes (4), after which the tank separates (5) and disintegrates over the Indian Ocean (6). The orbiter is now about 60 miles up and is traveling at Mach 15— fifteen times the speed of sound.

To prepare for reentry, the cargo-bay doors are closed. Then small yaw jets, clustered fore and aft, turn the orbiter tail first (9-11). An OMS deorbit burn (12), two to three minutes long, slows the craft by about 200 mph.

Pitch jets in the nose flip the orbiter over (13-14), angling the nose up 30 to 40 degrees (15) to protect the craft from temperatures as high as 2,800° F. At about 400,000 feet, the atmosphere begins to slow the orbiter's initial reentry speed of 16,500 mph and to turn its belly tiles bright orange (16). Ionized molecules in the air envelop the orbiter for about 13 minutes, cutting off radio contact.

Repeated S turns (17-21) slow the orbiter. The last reaction-control jets are shut off when the rudder becomes usable at 45,000 feet; the orbiter is flying at Mach 3.5 and falling 10,000 feet a minute with a glide angle of 22 degrees. At 1,725 feet, the nose pulls up, and the glide angle is reduced to 1.5 degrees by altitude 135 feet. Touchdown speed is about 185 knots (22).

F. Holz

SPACE SHUTTLE

The shuttle orbiter rests beneath the hoisting tower *(right)* at the Dryden Flight Research Center, Edwards Air Force Base, California. The massive truss is used to lift the orbiter so that NASA's Shuttle Carrier Aircraft (SCA) can roll under and hook up with it. The SCA—a modified Boeing 747 jumbo jet—then carries the orbiter piggyback *(below)* on test runs and on flights from the landing strip in California or New Mexico to the Cape Canaveral launch pad.

SPACE SHUTTLE

MISSION PORTFOLIO

The maiden voyage of the space shuttle *Columbia* marked an epoch in space travel, for *Columbia*—launched April 12, 1981—was rocket, spaceship and airplane all in one. Mission commander John Young and pilot Bob Crippen were the first astronauts who could bring their craft back from space to a dignified touchdown on land. Young and Crippen spent two days in orbit, flying upside down and backward, testing *Columbia's* computers, engines and cargo-bay doors. Then they pitched the craft over and began their gliding descent, nose up so that *Columbia's* tiled underbelly could dissipate reentry heat. Said Crippen as they crossed the coastline en route to Edwards Air Force Base: "What a way to come to California!"

Riding piggyback atop the external fuel tank, the *Columbia* orbiter lifts off from Pad 39A at Cape Canaveral at 7 a.m. on April 12, 1981. The shuttle's three main engines and

Trailing a plume of fire and dense smoke,
Columbia mounts skyward from the Florida
coast. The launch shook buildings three
miles away, yet felt "smooth," astronaut John
Young reported, "like riding a fast elevator."

away from *Columbia* 2 minutes 11 seconds after blast-off to parachute into the Atlantic 174,000 feet below. The rockets burned more than eight tons of propellant a second.

Columbia's 80-ton, 149-foot-long booster casings are towed back to Cape Canaveral to be refurbished and reused. The pair of steel casings came down on target, 2.5 miles apart, approximately 160 miles off Daytona Beach.

Columbia glides to a 215-mile-an-hour dead-stick landing at Edwards AFB, California, on April 14, 1981. The 102-ton orbiter touched down on a dry lake bed just 2 days 6 hours 20 minutes and 52 seconds after it blasted off.

Minutes after landing, *Columbia*—its crew still inside—waits as service vehicles approach to ventilate the craft and remove residual fuel from the engines. A wind machine *(far right)* blows away toxic fumes from *Columbia's* nose.

Still wearing flight coveralls, astronauts Crippen *(left)* and Young stand next to their wives after being flown back to Houston. Young called *Columbia* "a joy to fly. It does what you tell it to, even in very unstable regions."

On November 12, 1981, after battling an assortment of delaying problems, the world's first reusable spacecraft was finally reused. But the gremlins that had plagued *Columbia* on the ground pursued it into space. A faulty fuel cell, discovered on the second orbit, forced astronauts Joe Engle and Richard Truly into a "minimum mission" of just 54 hours.

Still, more than 90 per cent of the scheduled objectives were accomplished in less than half the scheduled time. Five remote sensing devices on an open pallet surveyed the earth to map terrain, measure air pollution, even study plankton in the sea. The multijointed Remote Manipulator System flexed and rotated to perfection. And *Columbia* itself, put through demanding reentry maneuvers, confirmed the first crew's high opinion. "The bird is real solid," said an impressed Engle. "A good solid bird all the way."

Columbia takes off from Cape Canaveral riding twin pillars of fire. The shuttle's second launch did less damage than the first, thanks to a newly installed water-deluge system that absorbed 75 per cent of the shock wave.

Astronauts Joe Engle *(left)* and Richard Truly grin at well-wishers as they get ready to board *Columbia* on the morning of launch. Truly, who turned 44 that day, said, "I'm going to have the biggest birthday candle I ever had."

Built in Canada (and proudly labeled so), the Remote Manipulator System crooks out of *Columbia's* open cargo bay toward earth, visible at the top of the picture. The 50-foot-long mechanical arm, designed to deploy and retrieve satellites, has two guidance cameras—one near the arm's elbow *(top)* and one near the wrist.

Showing the black heat-resistant tiles lining its belly, *Columbia* glides toward another perfect landing on November 14, 1981. Tiny white contrails stream from the orbiter's wing tips.

Escorted by a T-38 chase plane, *Columbia* touches down on the shimmering bed of Rogers Dry Lake at Edwards Air Force Base. *Columbia* had been up 54 hours 13 minutes and 12 seconds—7 minutes and 40 seconds less than it spent aloft on its maiden flight.

STS-3, as the third mission was called, began more routinely and ended more dramatically than either previous flight. In between, astronauts Jack Lousma and Gordon Fullerton had to overcome a number of exasperating problems—including broken cameras, a clogged toilet and fatigue—to accomplish a record schedule of experiments. They tested everything from *Columbia's* electromagnetic effect on space to space's effect on flying insects. They subjected the craft to prolonged doses of scorching sunshine. And, most important, they proved that the mechanical arm could, as promised, maneuver a payload.

Heavy rains in California had forced NASA to schedule the landing at White Sands, New Mexico. Then, during what was to have been the final orbit, high winds and a blinding sandstorm hit the new landing site, forcing a postponement of the landing. Finally, after eight full days aloft, and nearly 21 hours late, the orbiter touched down. The still-gusting winds gave it a landing speed of 250 mph—about 30 mph more than STS-1 or STS-2.

Columbia sits on Pad 39A at Cape Canaveral, Florida, prior to the launching of STS-3 on March 22, 1982. Leaving the huge external tank a reddish brown color—the hue of its spray-on foam insulation—saved the mission $15,000 and 600 pounds in unnecessary white latex paint.

High above the Pacific Ocean, the Remote Manipulator System reaches back across the open cargo bay with a device for measuring the charged particles and electromagnetic fields surrounding the orbiter. Radioed Gordon Fullerton after testing the $100-million arm, "I am very impressed with this piece of machinery."

Jack Lousma waves as he and Gordon Fullerton leave *Columbia* after their landing at White Sands, New Mexico, on March 30, 1982. The two astronauts had flown 129 earth orbits, traveled 3.9 million miles and been aloft for eight days and five minutes—more than three times as long as either STS-1 or STS-2.

Columbia took off on the shuttle's fourth flight on June 27, 1982, and came back on the Fourth of July. Independence Day was an appropriate time to return: The program of test runs was now successfully completed. *Columbia* was ready for full operational use, and NASA was no longer dependent on one-time-only spacecraft.

Astronauts Ken Mattingly and Henry Hartsfield continued the arm maneuvers and thermal tests begun on earlier missions and looked after NASA's first commercial clients. They monitored the separation of biological compounds for a drug firm, activated the first $10,000 "getaway special"—nine zero-gravity experiments designed by Utah State University students—and operated a Pentagon package that included a space sextant and infrared sensors for detecting exhaust from enemy missiles. Then they touched down at Edwards Air Force Base before a crowd of 500,000 people, including President Ronald Reagan and his wife Nancy, in *Columbia's* first landing on a concrete runway.

Scattering birds and billowing smoke, *Columbia* lifts off from Cape Canaveral at 11 a.m. on June 27, 1982—136 milliseconds ahead of schedule. The perfect launch was marred two minutes later when the parachutes on the solid rocket boosters failed to open, and the $32-million casings sank in 3,500 feet of water in the Atlantic.

STS-4 commander Ken Mattingly *(left)* and pilot Henry Hartsfield prepare to board *Columbia.* As a test of operational readiness—and as a way of cutting costs—the pair got less guidance from the ground than previous crews.

With an escort plane flying alongside, *Columbia* prepares to touch down on a concrete runway at Edwards Air Force Base, California. The orbiter used 9,660 feet of the runway's 15,000 feet and took a full minute to roll to a stop.

On November 11, 1982, less than eight hours into the fifth mission of the space shuttle, astronauts Joseph Allen and William Lenoir deployed SBS-3, a communications satellite owned by Satellite Business Systems, from *Columbia's* cargo bay. Both the men and their task were new. Allen and Lenoir were the first mission specialists to be flying on the shuttle, joining commander Vance Brand and pilot Robert Overmyer in the shuttle's first operational four-man crew. And the deployment of SBS-3 marked the first time an earth-orbiting satellite was ever launched from a piloted spacecraft. A day later the specialists made their record "two for two" by launching Canada's Anik C-3. Mechanical problems in their spacesuits prevented the astronauts from taking a space walk that had been scheduled; but deploying the satellites, at a combined cost to the owners of $17 million, justified the astronauts' boast that "we deliver"—and solidly launched NASA into the space trucking business.

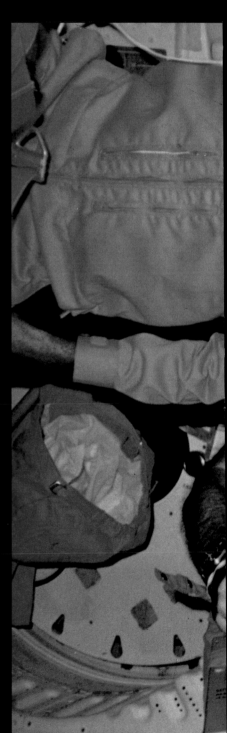

Trailing a cloud of its own, *Columbia* climbs high above the cloudy Atlantic after blasting off from Cape Canaveral. The veteran orbiter landed at Edwards Air Force Base five days later, then went into dry dock for improved engines, new instrument displays and additional seats.

The crew of the "Ace Moving Co." gathers for some shuttle advertising: pilot Overmyer upside down at top, commander Brand at center and mission specialists Lenoir *(left)* and Allen *(right)*. The president of Satellite Business Systems, first company to enjoy this "fast and courteous service," described *Columbia* as "a magnificent tool for the satellite business."

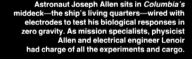

Astronaut Joseph Allen sits in *Columbia's* middeck—the ship's living quarters—wired with electrodes to test his biological responses in zero gravity. As mission specialists, physicist Allen and electrical engineer Lenoir had charge of all the experiments and cargo.

SATELLITE DEPLOYMENT BY
the Ace Moving Co.
FAST AND COURTEOUS SERVICE
"WE DELIVER"

In a series of photographs *(lower left to upper right)* taken from the aft windows of the flight deck, the SBS-3 communications satellite springs from its protective cradle in the orbiter's cargo bay at about three feet per second. Forty-five minutes later, when *Columbia* was 20 miles away, a rocket engine under the satellite ignited to blast it into geostationary orbit, parking it 22,300 miles above a spot on the Equator 840 miles west of Ecuador.

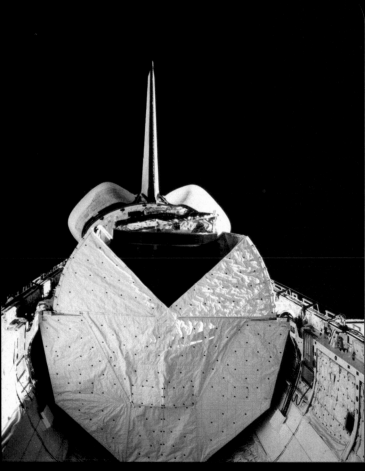

At Mission Control in Houston, flight director John T. Cox watches the clamshell-like thermal shield open to reveal SBS-3 before its deployment. Control of SBS-3 and Anik C-3 (*Anik* is ''little brother'' in Eskimo) passed to the satellites' owners immediately after deployment.

"We've got two veterans now," said one NASA official as *Challenger* began its long-delayed maiden voyage on April 4, 1983. Flown by commander Paul Weitz and pilot Karol Bobko, the new orbiter performed splendidly from the start, suffering less surface damage and fewer flight anomalies during its five-day mission than *Columbia* ever did. Mission specialists Story Musgrave and Donald Peterson added to the ship's laurels by successfully deploying the first tracking and data-relay satellite (TDRS) and testing new spacesuits and tools on a four-hour space walk.

But the cargo stole the show. Six hours into its journey, the TDRS started tumbling wildly out of control. Three hours of desperate signaling from the ground eventually separated the satellite from the second stage of its booster. More than a month later, ground-controlled firings of the satellite's thrusters began gently nudging the TDRS toward its intended orbit.

At 1:30 p.m. on April 4, 1983—the first afternoon launch for a shuttle flight—*Challenger* takes off from Cape Canaveral. NASA's second orbiter could generate more power and carry heavier loads than *Columbia*. The entire assembly had lost weight—roughly 10,000 pounds from the structure and mounts for the external tank alone.

Only the orbiter's wing tip remains visible as *Challenger* arcs over shortly after the flawless launch. "Boy, that was something," radioed mission commander Paul Weitz when the solid rocket boosters fell away two minutes after lift-off. "We recommend this heartily."

The $135-million tracking and data-relay
satellite lies in *Challenger's* cargo bay; its two
gold-colored-mesh dish antennas resemble
folded umbrellas. TDRS-A was the first of three
such satellites that were scheduled to
be deployed by the shuttle to provide global
communications coverage for the United
States and all of its satellites in earth orbit.

The TDRS seems to fall back toward earth in this photograph taken from *Challenger* shortly after the deployment. More than a month after communication with the TDRS had been reestablished, ground controllers began to boost the satellite little by little into its scheduled position. "We're not putting any time pressures on ourselves," one NASA official commented. "We're hurrying with restraint."

His tether hooked to a guide cable, Story
Musgrave moves along a cargo-bay handrail
during an EVA that he and Donald Peterson
performed on April 7, 1983. It was the first EVA of
the shuttle program and the first American
space walk since Skylab 3 in February 1974.

With *Challenger* in an inertial-hold attitude
177 miles above the coast of Mexico and the
Pacific Ocean, Musgrave *(above, left)* and
Peterson work in the aft cargo bay. During their
3-hour-52-minute EVA, the astronauts tested
zero-gravity tools and practiced procedures for
manually closing the cargo-bay doors.

Runway 22 at Edwards AFB lies dead ahead in this photograph taken by mission specialist Story Musgrave as *Challenger* makes its steep final approach on the 9th of April, 1983. Commander Paul Weitz—in manual control throughout the landing—brought *Challenger* to a halt on the runway midline with more than 9,000 feet to spare. ''It was a good burn,'' Weitz said of the reentry and landing, ''right down the pipe and smooth all the way.''

Soon after landing, astronauts Peterson, Bobko, Musgrave and Weitz *(from left to right)* get a warm welcome at NASA's Dryden Flight Research Center at Edwards. *Challenger's* first crew jokingly called themselves the ''Geritol Gang'': their average age—48 years 3 months —was the oldest of any NASA crew.

EXPLORING DEEP SPACE

While U.S. astronauts were traveling to the moon, repairing Skylab and learning how to fly the space shuttle, other envoys also were being dispatched beyond earth. Ungainly-looking but of remarkable ability, these explorers were a tribe of robots; their destination was deep space—the farthest reaches of our solar system and the planets that ply those uncharted seas.

Like the manned missions, the American program of robot probes was motivated in part by competition with the Soviet Union, whose first interplanetary shot in 1960—a failed attempt to reach Mars—preceded its American counterpart by two years. But the main stimulus for sending sophisticated machines millions of miles into space was science's need to know the answers to three tantalizing questions: What was the origin and evolution of the solar system? How did the universe come about? Does life exist elsewhere?

Attempting to answer these questions with deep-space probes placed special demands on planners and engineers. Because of the enormous distances to be traveled, and because each planet orbits the sun at a different speed, extreme accuracy and precise timing were required. Launch opportunities, or windows, occur only when the earth and the target planet are in the correct relative positions. A probe leaving earth may meet its target halfway around the sun from where it started. Moreover, since deep-space missions are measured in months and years, instead of the days and weeks of manned missions, the hardware had to be utterly reliable—and ground controllers extremely patient.

Although staff members at three different NASA centers have played prominent roles in planetary exploration, the prime mover among them was the Jet Propulsion Laboratory in Pasadena, California *(page 272)*. Early on, JPL took the lead in developing proposals for what a chronicler has called the "final far-out things" of planetary travel.

One of the ideas—the so-called Grand Tour—originated in 1963. The concept was based on an astronomical event that occurs only once every 175 years. Between 1976 and 1980, the five outer planets whose orbital paths around the sun lie beyond Mars—Jupiter, Saturn, Uranus, Neptune and Pluto—would come into an alignment highly favorable for exploration by a single spacecraft.

When a craft reached the vicinity of Jupiter, that planet's gravitational pull would deflect the robot on to Saturn. Employing this "slingshot effect" at each destination would enable a single spacecraft to tour as many as four outer planets while requiring only the original launch power needed to reach Jupiter.

Blacked out to show off its jewelry, Saturn wears not one ring, but more than 1,000 distinct ringlets in this composite photograph taken by *Voyager 1* in 1980. The gap in the rings on the right side was caused by Saturn's shadow.

Budget reductions and the emphasis on the space shuttle ultimately forced NASA to scale down its ambitious plans for planetary exploration, but throughout the 1960s, JPL developed techniques and hardware that made possible economy versions of the Grand Tour—and an exciting expansion of the space frontier.

The aims of JPL's first-generation Mariner probes were modest: to fly by a single target in the inner system of planets and send back information. Folded up and encased in a protective shroud during launch, a Mariner was put into earth orbit by a two-stage Atlas booster, then propelled on its journey by reignition of the second-stage engine. The probe had small jet thrusters to stabilize it in space and, on most missions, employed its own engine for midcourse corrections. Four winglike panels of solar cells powered the craft's radio and its array of scientific instruments.

To take full advantage of launch windows only a few weeks long, JPL typically sent its probes off in pairs. On July 22, 1962, *Mariner 1* blasted off for a fly-by of Venus. But 290 seconds into the mission, the craft veered off course and had to be destroyed by ground command. Five weeks later, however, *Mariner 2* lifted off flawlessly and on December 14 flew within 20,900 miles of Venus. The first news from another planet revealed that Venus was scorching hot. *Mariner 2's* radiometers measured surface temperatures of approximately 900° F.—more than four times as high as had been expected.

From inhospitable Venus, NASA turned to earth's neighbor away from the sun, Mars. *Mariner 3* failed when its protective shroud refused to separate after launch, but three weeks later, on November 28, 1964, *Mariner 4* began an eight-month journey to a fly-by 6,118 miles from Mars. The pay-off was 21 pictures, transmitted in digital form 134 million miles to earth at the agonizing rate of eight hours per picture. The images, the first ever made from near another planet, revealed no surface water—no trace of the Martian canals imagined by early astronomers. Instead, the Martian landscape was moonlike, pocked by craters.

The next Mariner probe returned to Venus. Launched on June 13, 1967, *Mariner 5* flew by that forbidding planet on October 19, just a day after the arrival of the Soviet craft *Venera 4,* which dropped a capsule onto the surface. Both probes confirmed the earlier temperature readings. Once thought to be almost earth's twin, Venus was, in the words of astronomer Carl Sagan, a "hellhole of a planet."

In 1969, two spacecraft were sent to visit Mars again—*Mariner 6* on February 24 and *Mariner 7* on March 27. For the first time, a pair of probes were bound for the same planet at the same time. They arrived within five days of each other and scanned a combined total of 201 pictures from a fly-by distance of 2,130 miles. The pictures were of higher resolution and much quicker transmission than *Mariner 4's* images—and also revealed an excitingly different aspect of the planet. In addition to the craters previously seen, there were flat desert-like regions and huge jumbled ridges and valleys where the craters had somehow been erased. The probes' infrared instruments sampled an atmosphere rich in carbon dioxide and suggested that the polar caps—long visible through earth telescopes—were frozen carbon dioxide, or dry ice.

Scientists wanted a closer look at this surprising planet. In 1971, conditions were ripe for putting a robot in

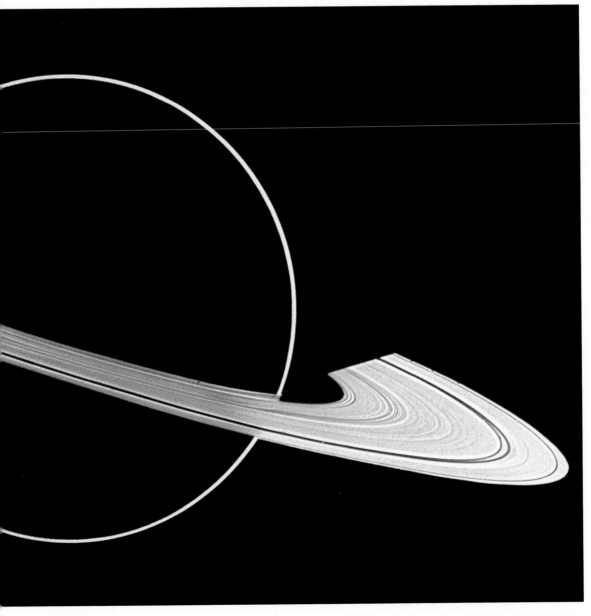

Martian orbit, but troubles plagued this complex project. On May 8, *Mariner 8* went awry less than five minutes after lift-off because of a second-stage engine failure. Then *Mariner 9*, launched on May 30, traveled more than five months and successfully entered Martian orbit—only to find the planet engulfed in a gargantuan dust storm that blotted out the probe's view of the surface.

Thanks to a more sophisticated on-board computer, however, ground controllers could reprogram *Mariner 9* in flight. The probe was ordered to hold off on its picture taking until further notice; when the dust settled, three months later, another signal was given to turn the cameras on.

Mariner 9—the first artificial moon for a planet beyond earth—orbited Mars for more than a year. Dipping to within 900 miles of the surface, it took 7,329 pictures and gathered other data that deepened the Martian mysteries. The overwhelming impression was of a planet with a turbulent geological history. Mars had dormant volcanoes, including a mammoth one that was three times the size of Hawaii's Mauna Loa, its biggest terrestrial competitor. The face of the planet was scarred with a four-mile-deep canyon that would stretch the breadth of the entire continental United States. Even more startling were numerous smaller channels, resembling terrestrial riverbeds, that appeared to have been gouged by the action of water. And spectrometer readings discovered traces of water in the atmosphere, with all of its implications for the presence of life.

While NASA planners prepared for Viking, an even more spectacular go at Mars, the last of the Mariner series was made ready for a new target. *Mariner 10's* goal was Mercury, the innermost planet and, because of its nearness to the sun, a particularly difficult one to study by telescope. To get *Mariner 10* to its destination, JPL planners made the first use of the gravity-assist technique—the sling-

shot originally intended for the Grand Tour of the outer solar system.

Mariner 10 left earth on November 3, 1973, and cruised to within 325 miles of Venus, sending back the first pictures from space of that planet. Then the gravity of Venus deflected the craft inward toward the sun and decreased its velocity by nearly one mile per second, putting it on the proper path to Mercury. A Teflon-coated sunshade opened to protect the probe from powerful solar rays and, on March 29, 1974—nearly 93 million miles from earth—the craft passed Mercury in a solar orbit that enabled it to fly by the planet twice more during the next year.

In 3,700 pictures and billions of bits of data, Mercury came through as a distinctly unfriendly place, its surface parched and cratered and its very thin atmosphere made up of traces of helium. But the planet was fascinating to scientists nonetheless, giving indications of a magnetic field and an unexpectedly high density—facts that pointed to the presence of an inner core of iron like that of earth.

The Mariner series was completed, but in 1975, the one-billion-dollar Viking exploration of Mars began. A joint project of JPL and NASA's Langley Research Center in Hampton, Virginia, the Viking spacecraft was designed to follow up Mariner's earlier findings of conditions conducive to life. Viking consisted of two parts: a

Mariner-like craft to orbit Mars and an instrument capsule designed to separate from the orbiter and descend to the Martian surface. The lander—sterilized to prevent contamination of Mars with earthly bacteria—would scoop up samples of Martian soil and a search for living microorganisms.

Viking 1 was launched on August 20, 1975, and its 7,750-pound twin, *Viking 2*, on September 9. The following June, *Viking 1* entered Martian orbit and began inspecting potential landing sites on the Plains of Chryse.

When the intended site appeared to be too rough, *Viking 1* was instructed to keep looking until it discovered something more suitable. Then the lander separated from the orbiter, fired an engine to brake out of orbit and deployed a parachute to slow its descent through the tenuous Martian atmosphere. One mile above the surface, it fired three retro rockets to slow descent even further, to 5 mph, and then settled itself gently onto the Martian surface.

It took 20 minutes for the news to travel 200 million miles from the Plains of Chryse to Southern California. When it finally arrived, Mission Control Center at JPL erupted with jubilation. It was July 20, 1976, seven years to the day after the first manned landing on the moon.

This robot and the *Viking 2* lander, which set down 14 days later on the

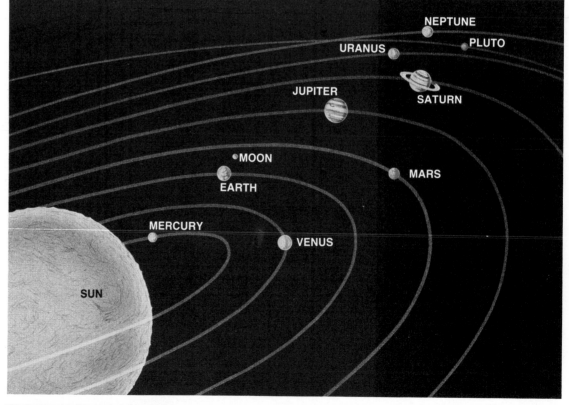

The planets and earth's moon are shown in their relative size to and order from the sun in this drawing used to map the voyages in this chapter. Neptune and Pluto cross orbits, jointly defining the solar system's outer edge.

Plains of Utopia to the north, were marvels of miniaturization and automation. No bigger than a jeep, each one contained three scientific laboratories and bristled with instruments, including two cameras and a 10-foot-long arm for collecting soil samples. The landers were powered by nuclear generators and controlled by their own computers.

Though the landers reported temperatures too frigid for most life—a high of −24° F. during the Martian summer—other readings seemed promising for the presence of some sort of microorganisms. The atmosphere contained 3 per cent nitrogen and may even have been warm and earthlike in an earlier time; ice made of water was found to underlie the dry ice of the polar caps.

The critical test would be the soil samples. On its eighth day on Mars, the *Viking 1* lander extended its arm, gouged a three-inch-wide trench and scooped up a batch of Martian dust. The sample was automatically conveyed back to the miniature laboratories where it was subjected to three different kinds of tests, each one intended to incubate Martian microorganisms, if they existed, and to detect chemical by-products of their activity such as carbon dioxide.

Initial results excited biologists. The soil samples contained traces of water, and the experiments showed a virtual rampage of chemical activity. At length, however, scientists concluded that this activity could be explained in terms of chemistry rather than biology. That conclusion—"undeniably disappointing," as one biologist said—was buttressed by a separate molecular-analysis experiment, which found no evidence of the organic compounds that constitute all forms of life on earth.

Though the apparent absence of life was discouraging, Viking's technological accomplishments were awe-inspiring. The landers and orbiters produced more than 50,000 images. Designed to operate for only 90 days, the *Viking 1* orbiter worked for four years before literally running out of gas—the propellant that powered its stabilizer jets. The *Viking 1* lander outlasted them all, sending weekly messages from the Martian surface until late 1982, when new instructions from the ground inadvertently wiped out part of the robot's computer memory.

NASA's first dazzling tour of the inner planets ended in 1978 with Pioneer-Venus, a double-barreled attempt by the Ames Research Center, Moffet Field, California, to penetrate Venus' dense veil of clouds. The hardware for *Pioneer-Venus 1,* the orbiter, was descended from earlier generations of Pioneer craft designed by JPL to study cosmic radiation and other interplanetary phenomena.

Pioneer-Venus 1 was launched on May 20, 1978, and went into orbit around Venus on December 4. It took ultraviolet pictures of the clouds and penetrated them with radar beams that measured elevations of the planet's surface for the first time, revealing canyons and continent-sized plateaus, along with what looked like craters and dormant volcanoes.

Pioneer-Venus 2, launched on the 8th of August, was a different kind of spacecraft. As it approached Venus, the vehicle separated into four cone-shaped probes that sampled the dense atmosphere as they plunged to the planet's hidden surface. One important finding confirmed earlier suggestions of why Venus is so hot: Carbon dioxide in the atmosphere acts like the glass in a greenhouse, trapping solar radiation and overheating the surface.

Even before this probe of Venus, the Ames Center had used another version of the Pioneer spacecraft to undertake the first reconnaissance of the planets that lie beyond Mars. *Pioneer 10,* launched on March 2, 1972, passed the orbit of Mars and flew to within 82,000 miles of Jupiter after a flight of 21 months. Its partner, *Pioneer 11,* launched on April 5, 1973, accomplished the fly-by of Jupiter in December 1974. Then, given a Jovian gravity assist, *Pioneer 11* was flung into nearly five years of additional travels, to pass by Saturn on the 1st of September, 1979, at a distance of 13,300 miles.

Impressive in their own right, these two Pioneers, which transmitted more than 400 images, were intended principally as pathfinders. Their adventures in braving the asteroid belt of rocky debris between Mars and Jupiter, as well as the powerful Jovian radiation fields—10,000 times as strong as earth's Van Allen belts—influenced the design of an even more ambitious follow-up project known as Voyager.

At a cost of $780 million, Voyager—managed jointly by JPL and Langley Research Center—was the scaled-down version of the spacecraft NASA had once envisioned for the Grand Tour of the outer planets. The 1,797-pound Voyager made use of the slingshot technique and parts of the hardware intended for the original probe, including new nuclear power packs and self-correcting computers. Because of the nearly three hours required for Voyager data to reach earth and for new instructions to make the return trip, these computers were programed to react to emergencies on their own.

Voyager 1 was launched on September 5, 1977, actually 16 days after *Voyager 2.* It took a shorter trajectory and thus arrived in the vicinity of Jupiter on March 5, 1979, four months ahead of *Voyager 2.* Among the early images from the total of more than 68,000 pictures transmitted by the two Voyagers was a striking view of gas and debris spewing from a volcano on the largest Jovian moon, Io. It was the first discovery of an active volcano beyond earth.

With a boost from Jupiter's gravity, the twin Voyagers then hurtled on toward Saturn—a journey of 20 months for *Voyager 1* and more than two years for *Voyager 2.* As both Voyagers viewed it, Saturn was a cold gaseous ball racked by enormous whirling storms driven by winds of up to 1,100 mph. Saturn's half-dozen rings appeared to be composed of thousands of smaller ringlets, probably consisting of icy bits of cosmic debris. And the Voyagers discovered six new moons, bringing to 23 the number of satellites known to be orbiting Saturn.

Its job finished, *Voyager 1* kept going, bound for interstellar space. But the work of *Voyager 2* had only begun. The probe received a gravity assist from Saturn and went into a new trajectory intended to carry it close to Uranus in 1986 and Neptune in 1989—giving NASA its Grand Tour after all. Both probes are expected to keep transmitting for several years.

Meanwhile, NASA's shrinking budget permitted the scheduling of only one major new deep-space probe for the 1980s. A spacecraft called Galileo, designed to be carried into earth orbit by the space shuttle, would then be launched by a Centaur rocket toward Jupiter—to orbit that giant planet and to release a probe into its dense atmosphere.

Yet even as planners prepared for Galileo and scientists sorted out the voluminous data already gleaned from the Voyagers' first epic reconnaissance of the solar system, a frail envoy from earth was sailing beyond all that. After its encounter with Jupiter in 1973, *Pioneer 10* had traveled on. On June 13, 1983, it crossed the orbit of Neptune, more than 2.8 billion miles from the sun, and thereby became the first human artifact to leave our solar system.

In the event that *Pioneer 10* should encounter intelligent life in the vastness of interstellar space, it carries a message. Affixed to the spacecraft is a six-by-nine-inch aluminum plaque engraved with symbols that indicate the location of our planet and with images of a man and a woman. The man's right hand is raised in a gesture of earthly good will. □

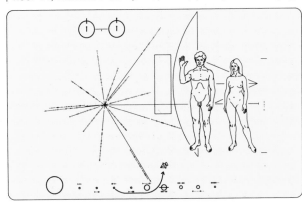

A man and woman stand before a spacecraft on this plaque, a cosmic greeting card carried by *Pioneer 10*. The radial pattern situates the sun within our galaxy; the sun and planets, with a satellite leaving earth, are depicted at bottom.

BASIC SPACE PROBES

In the saga of interplanetary exploration, space probes have been hardware heroes. Acting as extraterrestrial laboratories for earthbound scientists, American or Soviet, the probes carry a complex of sophisticated equipment—infrared radiometers to measure a planet's surface temperature, ultraviolet spectrometers to analyze the composition of its atmosphere, magnetometers to detect its magnetic field, however slight. The rest of a probe's components exist solely to serve its scientific heart: to power it, to protect it and to relay its findings back to earth.

Once launched on its trajectory beyond the reach of earth's gravitational pull, a space probe's navigation system keeps it on course. In most cases, this job is performed by tracking sensors and the onboard computer system. With one sensor fixed on the sun and another on a celestial body such as the star Canopus, the computer can calculate the craft's attitude and position, then make necessary corrections by activating gas-powered thruster jets.

In some space probes, such as the Mars-orbiting *Mariner 9 (below)*, more significant course corrections are made by a liquid-fuel rocket engine. In *Mariner 9*, these course corrections were initiated by controllers on earth. More advanced computer systems, such as that of *Pioneer 10 (opposite, below)*, are programed to make major course corrections on their own, but if the need arises, the system can be reprogramed by ground controllers.

Another critical part of every space probe is the communications system, which transmits and receives radio signals across the vast reaches of space. All probes carry low- and medium-gain antennas for this purpose; probes to the outer planets also carry large dish-shaped antennas to amplify the signals that they send back to earth.

In order for all of these electronic systems to function, the probe must also be equipped with a power source. Probes exploring the inner planets of the solar system, from Mercury to Mars, can draw on earth's oldest and greatest of power sources: the sun. Attached to the body of all these spacecraft are rectangular panels containing thousands of photovoltaic cells to convert solar energy into electrical current. Probes headed toward the sun require only two solar panels to produce the necessary current; those outward bound to Mars require four.

Probes such as *Pioneer 10*, designed to travel still farther from the sun, cannot depend on sunlight as a power source. Instead of solar panels, these craft carry small nuclear generators on outriggers. To prevent the radiation produced by the generators from skewing the probe's experimental findings, scientific instruments and the generators must be positioned as far apart as possible.

Protecting all these systems from the extremes of the interplanetary environment are temperature-control systems. Insulating blankets, ventilation louvers and heat shields are all designed to prevent the probes from overheating as they travel toward the sun. Those outward bound are equipped with heaters to prevent them from freezing as they whir and click away in the vacuum of space. □

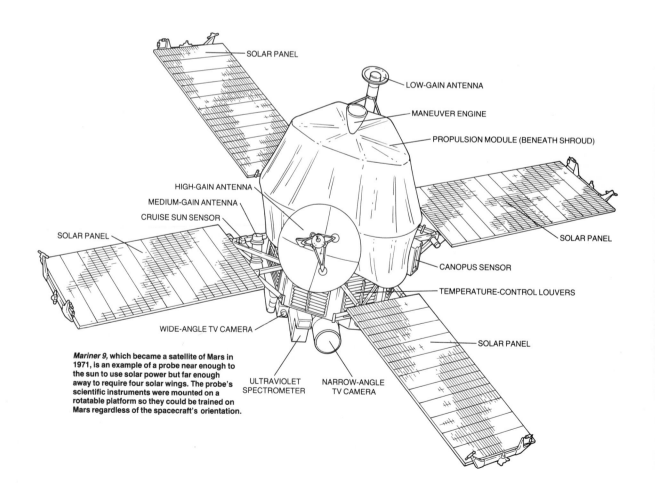

Mariner 9, which became a satellite of Mars in 1971, is an example of a probe near enough to the sun to use solar power but far enough away to require four solar wings. The probe's scientific instruments were mounted on a rotatable platform so they could be trained on Mars regardless of the spacecraft's orientation.

SOLAR PANEL

LOW-GAIN ANTENNA

MANEUVER ENGINE

PROPULSION MODULE (BENEATH SHROUD)

HIGH-GAIN ANTENNA

MEDIUM-GAIN ANTENNA

CRUISE SUN SENSOR

SOLAR PANEL

SOLAR PANEL

CANOPUS SENSOR

TEMPERATURE-CONTROL LOUVERS

WIDE-ANGLE TV CAMERA

SOLAR PANEL

ULTRAVIOLET SPECTROMETER

NARROW-ANGLE TV CAMERA

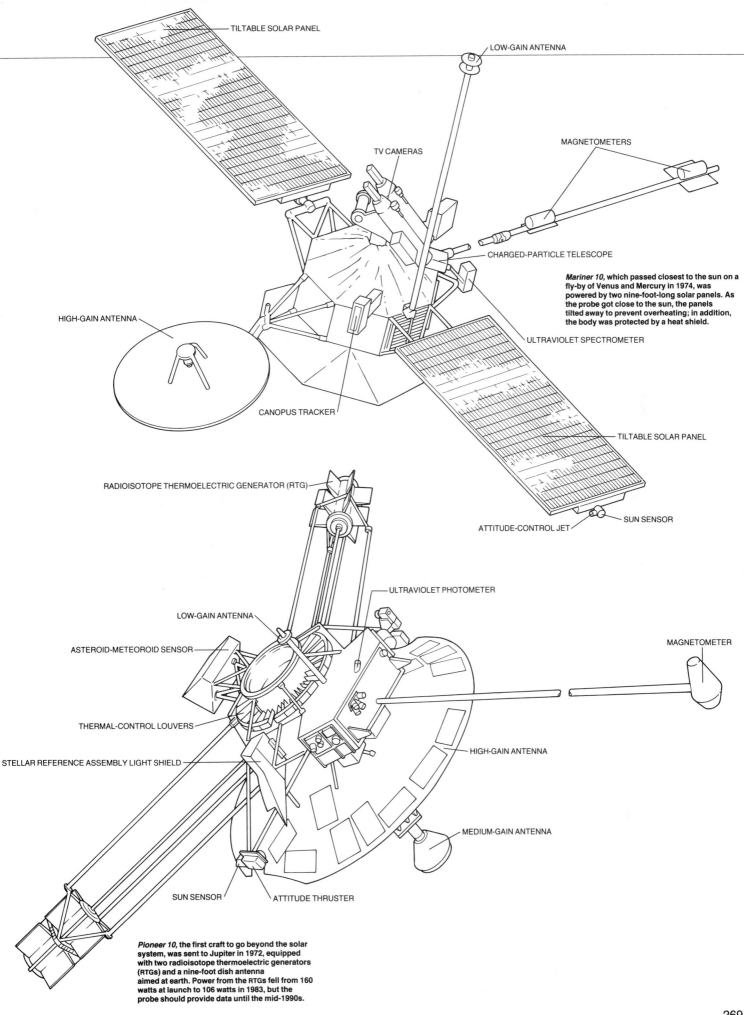

TILTABLE SOLAR PANEL

LOW-GAIN ANTENNA

TV CAMERAS

MAGNETOMETERS

CHARGED-PARTICLE TELESCOPE

HIGH-GAIN ANTENNA

Mariner 10, which passed closest to the sun on a fly-by of Venus and Mercury in 1974, was powered by two nine-foot-long solar panels. As the probe got close to the sun, the panels tilted away to prevent overheating; in addition, the body was protected by a heat shield.

ULTRAVIOLET SPECTROMETER

CANOPUS TRACKER

TILTABLE SOLAR PANEL

SUN SENSOR

ATTITUDE-CONTROL JET

RADIOISOTOPE THERMOELECTRIC GENERATOR (RTG)

ULTRAVIOLET PHOTOMETER

MAGNETOMETER

LOW-GAIN ANTENNA

ASTEROID-METEOROID SENSOR

THERMAL-CONTROL LOUVERS

HIGH-GAIN ANTENNA

STELLAR REFERENCE ASSEMBLY LIGHT SHIELD

MEDIUM-GAIN ANTENNA

SUN SENSOR

ATTITUDE THRUSTER

Pioneer 10, the first craft to go beyond the solar system, was sent to Jupiter in 1972, equipped with two radioisotope thermoelectric generators (RTGs) and a nine-foot dish antenna aimed at earth. Power from the RTGs fell from 160 watts at launch to 106 watts in 1983, but the probe should provide data until the mid-1990s.

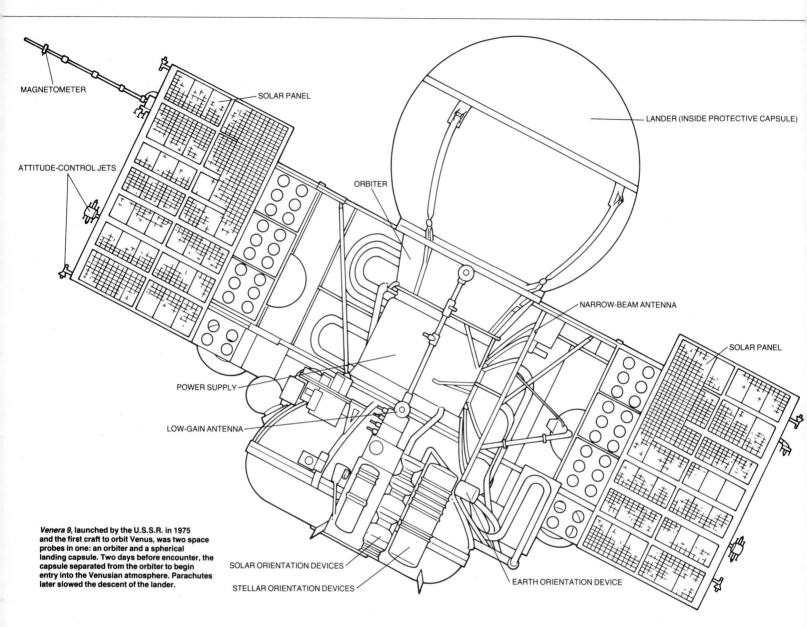

MAGNETOMETER

SOLAR PANEL

LANDER (INSIDE PROTECTIVE CAPSULE)

ATTITUDE-CONTROL JETS

ORBITER

NARROW-BEAM ANTENNA

SOLAR PANEL

POWER SUPPLY

LOW-GAIN ANTENNA

SOLAR ORIENTATION DEVICES

STELLAR ORIENTATION DEVICES

EARTH ORIENTATION DEVICE

Venera 9, launched by the U.S.S.R. in 1975 and the first craft to orbit Venus, was two space probes in one: an orbiter and a spherical landing capsule. Two days before encounter, the capsule separated from the orbiter to begin entry into the Venusian atmosphere. Parachutes later slowed the descent of the lander.

The stony wastes of the torrid Venusian landscape are revealed in this panoramic view photographed by the *Venera 9* lander, the first space probe to photograph the surface of Venus. The lander transmitted 53 minutes of data that were recorded and later relayed to earth by the orbiter section of the spacecraft.

LUNAR ROBOTS

Among the early unmanned probes to leave earth was a remarkable family of lunar explorers—the Surveyors. Between 1966 and 1968, five of these small, three-legged spacecraft landed gently on the surface of the moon to televise potential Apollo landing sites. They analyzed the lunar surface to determine whether it would bear the weight of a manned lunar module.

The Soviet Union employed a similar series of soft landings, beginning with *Luna 9* in 1966, to carry out experiments on the moon using robots instead of human explorers. In 1970, *Luna 17* disembarked the first moon car, a strange-looking eight-wheeled vehicle named *Lunokhod*—Russian for *Moon Rover*—that required a crew of five controllers back on earth to manage its assorted snail-paced movements.

Surveyor rests on its footpads on earth.

These two drawings depict the *Lunokhod* moon car rolling off a Luna spacecraft on ramps that are folded in flight. Its eight wire wheels, each with its own small electric motor, carried *Lunokhod* 6.5 miles in nearly 11 months before its miniature atomic power plant gave out.

JPL, DEEP SPACE HQ

In the summer of 1964, JPL, the Jet Propulsion Laboratory in Pasadena, California, was approaching a crisis. The installation, NASA's principal contractor for unmanned deep-space exploration, was having great difficulty getting a functioning spacecraft to the moon.

Four and half years earlier, NASA had commissioned JPL to produce a series of lunar probes for a project called Ranger. Launched in 1961 and 1962, the first five Ranger craft failed miserably. Only one of them, *Ranger 4,* actually reached the moon—but then the master clock in its central computer malfunctioned; *Ranger 4* could not perform timed automatic functions, nor could it respond to commands from earth. "All we've got is an idiot with a radio signal," said one NASA official.

After *Ranger 5* was derailed by a series of short circuits—and missed the moon by 450 miles—future launches were postponed and NASA subjected JPL to a rigorous review. The laboratory's management was drastically reorganized to improve systems-design review and ensure strict quality control. When *Ranger 6* was finally launched on January 30, 1964, it was the most thoroughly tested spacecraft built up to that time. It flew flawlessly and landed on target, in the Julius Caesar region in the Sea of Tranquillity. But its television cameras, practically the only scientific equipment on board, failed to transmit a single picture.

The cruel disappointment of *Ranger 6* prompted a Congressional investigation and still tighter NASA supervision of JPL. The press—and the comic strips as well—took pot shots at the laboratory. Comic-strip detective Dick Tracy handily solved the mystery of the television black-out while on a moon cruise in a friend's space coupe. Spotting the wreckage of *Ranger 6,* he stepped out to investigate. "It's easy to see why the cameras failed," Tracy said. "They were never turned on."

The next summer, as the launch of *Ranger 7* approached, pressure on JPL scientists was enormous. But even during the countdown, Director William H. Pickering never lost his good humor. When someone reported a high noise level in the booster-rocket guidance system, Dr. Pickering had a ready answer. "That," he said, "is the noise from Washington."

As it turned out, even Washington was happy with *Ranger 7.* It was, as Pickering put it, "a textbook operation—everything on the button." *Ranger 7's* cameras worked perfectly, transmitting 4,316 pictures—all of them better than previous earthly images of the moon's surface by what one astronomer called a factor of 1,000. As the walls of JPL's control center rang with cheers, someone asked Project Manager Harris M. (Bud) Schurmeier if *Ranger 7* had

disintegrated on impact. "We'll get Dick Tracy to check on this one also," Schurmeier replied.

Ranger 7's success established JPL's preeminence in space exploration. The laboratory had begun in the mid-1930s as an informal rocketry adjunct to the California Institute of Technology's Guggenheim Aeronautical Laboratory. That is, it consisted of a few graduate students experimenting with rocket engines on a dry stream bed near the CALTECH campus. During World War II, the rocket project won Army contracts and took the name Jet Propulsion Laboratory—the term "rocket" was thought to be too closely associated with Buck Rogers space fantasies.

Through the 1940s and 1950s, JPL developed a few early tactical guided missiles as well as rocket engines to boost heavily laden aircraft during

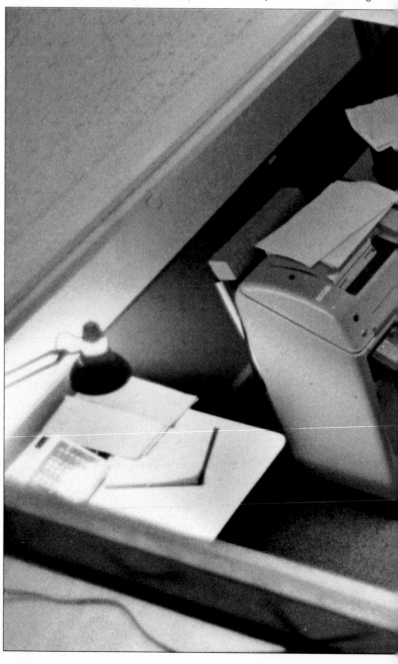

Project Manager Bud Schurmeier passes agonizing hours in a cubicle at the JPL command center in Pasadena as *Ranger 7* hurtles toward the moon. Six previous failures had left the laboratory's future riding on the mission.

JPL engineers in the command center anxiously watch the monitors that report *Ranger 7's* progress as it nears the moon. Journalists were not permitted in the room; the photograph was taken with a remote-control camera.

Jubilant staffers prepare to toast the success of *Ranger 7* just moments after the spacecraft landed on the moon. "For those of us who had lived Ranger for so long," recalled one, "it was kind of a spiritual happening."

JPL Director Pickering and Project Manager Schurmeier point out *Ranger 7's* landing site at a press briefing after the mission. Asked about JPL's once-clouded future, Pickering answered with a grin, "I think it's improved."

Engineers tend computer terminals in JPL's
control center during *Voyager 1's* encounter with
Saturn in November of 1980. The two large
images behind them on the left indicate that the
center is in two-way communication with both
Voyagers through a tracking station in Australia.

takeoff. But by 1957, even before the orbiting of *Sputnik I,* the laboratory had determined that its future lay not in weapons but in space. Together with the U.S. Army Ballistic Missile Agency, JPL was responsible for the successful launch of *Explorer I,* the first American earth satellite, in January 1958. At the end of that year JPL was taken under the wing of the newly created NASA, and plans were laid for the Ranger moon shots and the Mariner planetary probes to Venus, Mars and later to Mercury.

After the success of *Ranger 7* came the JPL-supervised Surveyor moon shots, which made the United States' first controlled, soft landings on the lunar surface. In the next decade, JPL built the Viking Mars orbiter and launched the twin Voyagers bound for Jupiter, Saturn and beyond. By 1980, JPL had a staff of 4,600 and a budget of $400 million. The pace of planetary exploration has since slackened somewhat, but the laboratory in Pasadena remains the principal headquarters for what retired Director Pickering once described as "things right on the fringe of impossibility." □

The Jet Propulsion Laboratory's 177-acre "campus" gleams in the brilliant sun beneath the San Gabriel Mountains on a rare smogless day in Pasadena, California. The wide, dark-windowed building at upper left houses the laboratory's administrative headquarters.

DEEP SPACE

MISSION PORTFOLIO

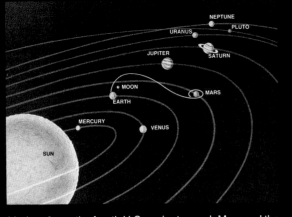

Mariner 9 was the fourth U.S. probe to reach Mars and the first spacecraft of any kind to orbit another planet. It was launched from the Kennedy Space Center on May 30, 1971, entering its highly elliptical, 12-hour Martian orbit 167 days later. Despite being greeted by a tremendous dust storm, the spacecraft radioed back tens of billions of bits of information and photographed fully 90 per cent of Mars's surface as well as both its tiny moons.

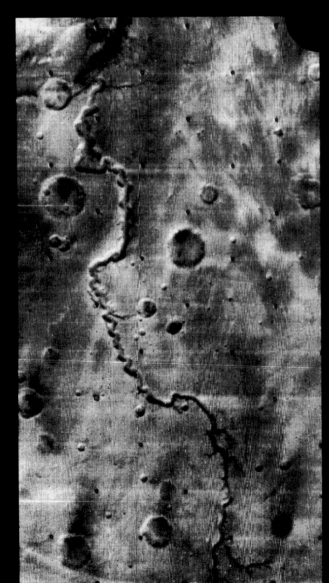

This meandering, 250-mile-long valley is as close as Mariner 9 came to finding the Martian "canals" of popular legend. Only water could have dredged such a channel on earth—a fact that suggests the possibility that Mars, nearly bone-dry at present, may once have had rivers.

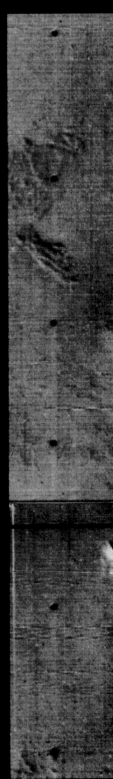

In a mosaic of four images relayed by Mariner 9 in February 1972, Olympus Mons, Mars's own Mount Olympus, overflows an outline of the state of Missouri superimposed to give a sense of scale. The immense volcano—three times as large as any earthly one—is 17 miles high and 375 miles across at its base. Ancient lava flows run down to the surrounding plains.

SUN

Launched in November 1973, *Mariner 10* was the first probe targeted for Mercury, roughly two thirds of the way from earth to the sun. En route it became the third U.S. probe to study Venus, whose gravity helped it alter course and pick up speed. It passed within 431 miles of Mercury in March 1974, then entered a solar orbit that gave it another Mercury fly-by in September and a third the next March.

Venus appears deceptively cool in this ultraviolet photograph taken from 450,000 miles away by *Mariner 10* on February 6, 1974, as the probe passed by on its way to Mercury. The blue tint is the result of a filter used to enhance cloud outline and reveal the circulation pattern in the dense atmosphere of Venus.

photomosaic, composed of 18 pictures taken
from 124,000 miles out during *Mariner 10's* initial
approach on the 29th of March, 1974.

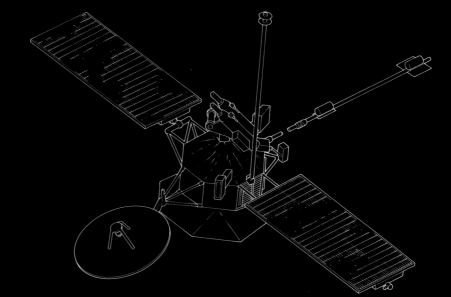

Mariner 10 seems to float like a cosmic glider,
equipped with solar panels for wings and a 20-
foot-long magnetometer boom for a tail.
The probe had 170 pounds of instruments and
weighed about 1,100 pounds in all.

A 360-mile section of Mercury's pocked, airless
surface fills this *Mariner 10* photograph taken
from 48,000 miles away. The jagged line at top
center is a long, high cliff probably created as
Mercury cooled and shrank; the apparent tear at
the top left is caused by a loss of image data.

Two true pioneers made the first journey through the asteroid belt to Jupiter: *Pioneer 10* completed its 21-month voyage on December 3, 1973, and *Pioneer 11*—taking a month less en route—arrived one year later. *Pioneer 10* continued on to cross Neptune's orbit in June 1983, thus becoming the first craft to leave the solar system. *Pioneer 11*—which will leave the solar system in another direction—made the first fly-by of Saturn in September of 1979.

Jupiter's eye-shaped Great Red Spot and the shadow of the moon Io—roughly the size of earth's moon—mark this *Pioneer 10* picture *(top)* taken from about 1.6 million miles above the cloud-ringed planet. *Pioneer 11* took its nearest photograph of the Red Spot *(right)* from 338,000 miles off. Scientists believe the spot may be a center of hurricane-like activity within the complex weather system on Jupiter.

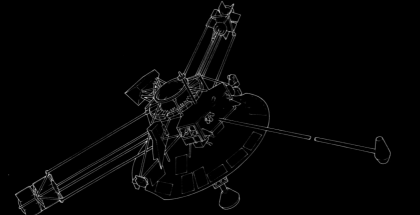

The probe designed for both *Pioneer 10* and
11 is dominated by its high-gain dish antenna—a
configuration that was made necessary by
the astronomical distances its transmissions
must travel to earth. Spindly outriggers
hold the nuclear generators that power the craft.

In this photograph taken from a distance of
1.8 million miles, Saturn's usually bright rings,
illuminated from below, appear as dark bands.
Pioneer 11 plunged through the ring plane
on the 1st of September, 1979, after traveling two
billion miles over six and one half years.

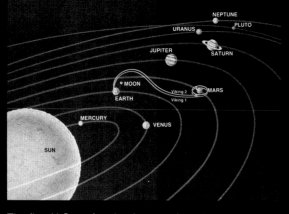

The first U.S. craft to land on another planet were aptly named after early explorers. *Viking 1* was launched for Mars on August 20, 1975, and *Viking 2* on September 9. Each had an orbiter to study the atmosphere, map the surface and relay data, and a lander to take atmospheric readings, analyze soil and search for life. After separating from the orbiter, the *Viking 1* lander touched down July 20, 1976; its twin landed 4,600 miles away two weeks later.

Thirty-eight images recorded by the *Viking 2* orbiter as it passed over the Martian south pole make up this photomosaic of Mars's southern latitudes. Taken at the time of the planet's vernal equinox in 1977, the pictures show winter frost receding toward the pole—just out of sight below the bottom center photographs.

The 5,125-pound Viking orbiter carries a TV camera and other instruments on a platform underneath the main body. A tiny antenna near the end of one solar panel *(lower left)* relays messages to and from the lander *(overleaf)*, while the larger high-gain antenna maintains two-way communication with earth.

Reddish ground and a salmon-colored sky
stretch away from the *Viking 2* lander on the
Plains of Utopia near Mars's north pole.
The colors—identical to those found by *Viking 1*
—reflect the heavy concentration of iron
in the soil and of dust in the atmosphere. From
lander to rocky horizon is about two miles.

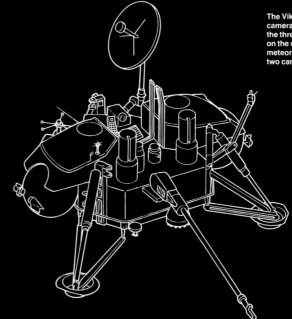

The Viking lander bristles with antennas, cameras and probes. Extending between two of the three legs is the surface sampler boom; on the right side, a short boom holds meteorology sensors. Flanking the boom are two canisters that house television cameras.

With only its weather boom visible, the *Viking 1* lander surveys a field of dunes on Mars that looks—at least in black and white—as though it might be in Mexico or California. This scene was photographed two hours after sunrise on August 3, 1976, with the sun 30 degrees above the horizon at right. The large boulder at left measures approximately 3 by 10 feet.

Olympus Mons towers above midmorning clouds in this artist's rendering *(right)* of a black-and-white image *(below)* made by the *Viking 1* orbiter approximately six years after *Mariner 9's* discovery of the enormous volcano. Taken from 5,000 miles away, the photograph reveals a multiringed crater 50 miles wide. A distinct wave cloud train *(upper left)* extends several hundred miles beyond the mountain.

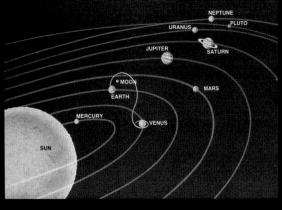

Pioneer-Venus 1 flew by its target on December 4, 1978, then settled into a highly eccentric orbit that brought it within 113 miles of the Venusian surface once every 24 hours. Special ultraviolet equipment returned images of the turbulent atmosphere, while a radar altimeter penetrated the clouds to create the first contour map of the surface.

Venus' atmospheric cloak shines gold in this unfiltered ultraviolet picture taken by *Pioneer-Venus 1* from 40,000 miles away. The dark strands are storm systems racing around the planet at 220 mph; the thick clouds are made up of suspended particles of sulfuric acid.

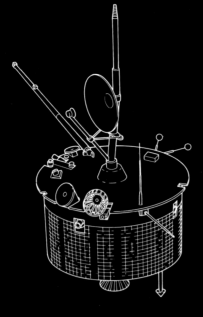

The drum-shaped Pioneer-Venus orbiter—eight feet across, with solar cells around its sides—carries four antennas, a magnetometer and other scientific instruments. The orbiter is expected to relay data through 1986, then drop into the Venusian atmosphere in 1992.

Blue lowlands, green plateaus and red peaks blister Venus' surface in this color-enhanced computer reconstruction made from data collected by *Pioneer-Venus 1*. Two "continents" are visible: Ishtar, top, and Aphrodite, bottom right. The pole's blank cap signifies an absence of data; the area was in the probe's blind spot.

In the late summer of 1977, twin Voyagers set off for a closer look at the two planetary giants, Jupiter and Saturn, and their many satellites. *Voyager 1* braved the intense Jovian radiation belts to pass 172,000 miles above Jupiter in March 1979. *Voyager 2,* while playing it safer with Jupiter, flew within 63,000 miles of Saturn on August 25, 1981, then headed on to keep humanity's first dates with Uranus and Neptune in the late 1980s. That done, only Pluto, of all the planets, will remain unvisited.

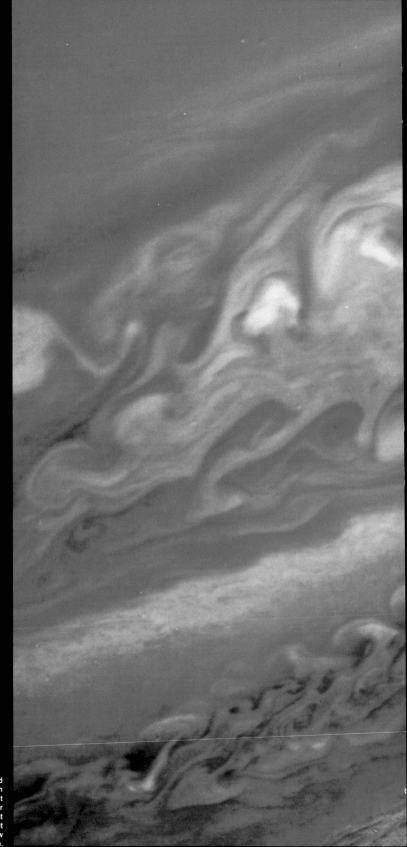

The Voyager space probe consists of a great "flying dish"—a 12-foot-wide high-gain antenna aimed at earth—attached to a 10-sided mission module spiked with antennas and instrument booms. The boom at lower left holds three nuclear generators, which produced 420 watts of power at launch, falling to 380 after encounter with Saturn.

Cloud bands and lesser storms surround the Great Red Spot of Jupiter in this photograph from *Voyager 1.* Scientists speculated about the spot for more than 300 years before Voyager proved it to be a monumental storm that endures through a combination of size—it stretches 20,000 miles end to end—low temperature and thick surrounding atmosphere.

Two of Jupiter's moons—Io *(at near right, above the Great Red Spot)* and Europa *(at far right)*—stand out against the giant planet in this photograph assembled from three images taken by *Voyager 1* on February 13, 1979. The probe was still more than 12 million miles away.

The Jovian moon Callisto—possibly the most densely cratered body in the solar system —bears the scars of ancient bombardments in its ice-and-rock crust. The ringed bright spot at the top measures about 200 miles across.

As seen by *Voyager 2*, bright spots mark the most recent impact craters on Ganymede, whose radius of 1,640 miles makes it the solar system's largest satellite—slightly bigger than the planet Mercury. Surface grooves suggest shifting among parts of Ganymede's icy crust.

Mazelike streaks cover the almost-craterless
surface of Europa—the smallest, the brightest
and probably the youngest of the four large
Jovian moons (Ganymede, Callisto, Io, Europa)
discovered by Galileo in 1610. The veins may
be shallow fracture lines filled with frozen water.

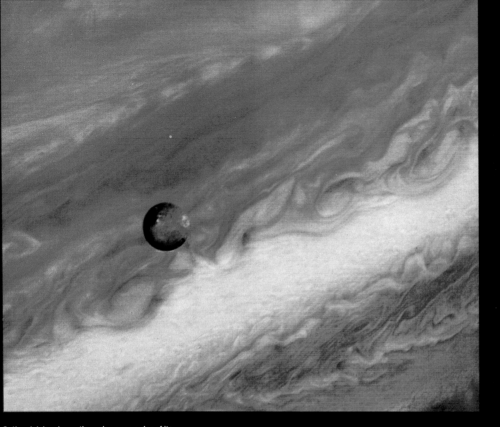

Sulfur-rich Io mirrors the red-orange color of its mother planet in this photomosaic taken by *Voyager 1*. Innermost of Jupiter's four Galilean satellites, Io revealed a striking topography, complete with mountains, cliffs and at least nine active volcanoes—but no impact craters.

A volcano spews fire and debris more than 100 miles above the bright rim of Io in this image made from 304,000 miles away by *Voyager 1* on March 4, 1979. Io is the most volcanic body in the solar system, emitting an average of two watts of energy per square meter over its surface —equal only to earth's most active regions.

Saturn's thousands of rings, ringlets and microringlets look like grooves in a psychedelic phonograph record in this color-enhanced image made by *Voyager 2* on August 17, 1981. The spacecraft took approximately 15,000 photographs as it swept to within 745 miles of the planet's 620,000-mile-wide halo, which proved only 500 feet thick in spots. The rings are composed of whirling rock-and-ice particles; their outer reaches orbit Saturn every 12 hours.

Second only to Jupiter in size, Saturn wears
brilliant equatorial rings and delicately shaded
bands of cloud in this image from *Voyager 1*.
The probe reached a peak speed of 56,600 mph
on its 1.35-billion-mile journey to Saturn
—and arrived a mere 12 miles off course.

Of all the worlds visited by NASA's interplanetary probes, the one that has turned out to be most like earth is Titan, Saturn's largest moon, shown here rimmed with a layer of bluish smog in a photograph taken by *Voyager 2* from 2.7 million miles away. One third the size of earth, Titan is the only moon with an atmosphere— one that, like earth's, consists mainly of nitrogen. Scientists believe that Titan is similar to earth three billion years ago—and that it may help answer questions about the chemical origins of life. They have called Titan "the most exciting spot in the solar system."

PICTURE CREDITS

The sources for the illustrations that appear in this book are listed below. Credits from left to right are separated by semicolons, from top to bottom by dashes. Dust Jacket: NASA. 6-9: Ralph Morse for *Life*. 12, 13: Robert Kelley for *Life*—Robert Williams for *The Commercial Appeal*; Frank Williams for *The Detroit Free Press*; John Bryson for *Life*. 14: Drawing by Fred Holz, based on original painting by Ray Pioch. 17: Courtesy Time Inc. Picture Collection—from *V-2* by Walter Dornberger © 1954 by Viking Press, Inc. Copyright renewed 1982 by Viking Penguin Inc. Reprinted by permission of Viking Penguin Inc.—Wide World; UPI (2). 18-21: Walter Sanders for *Life*. 22-29: Ralph Morse for *Life*. 30: Ralph Morse for *Life*—NASA. 31: Ralph Morse for *Life*. 32: Ralph Morse for *Life*. 33: Hank Walker for *Life*—Burris Jenkins Jr. for *The New York Journal-American*, courtesy Time Inc. Picture Collection; U.S. NAVY/NASA. 34, 35: Ralph Crane for *Life*. 36: NASA—Wide World—courtesy Time Inc. Picture Collection. 37: Wide World. 38-40: James Whitmore for *Life*. 42, 43: Ralph Morse for *Life* (2); NASA (3). 44-47: NASA. 48: U.S. NAVY/NASA—NASA from Wide World. 49: NASA. 50, 51: James Whitmore for *Life* (2)—Ralph Morse for *Life*. 52, 53: Michael Rougier for *Life*. 54: Michael Rougier for *Life*. 55: Michael Rougier for *Life*; Lynn Pelham for *Life*. 56: Ralph Morse for *Life*. 57: Michael Rougier for *Life*. 58, 59: Ralph Morse for *Life*; Al Fenn for *Life*; Bob Gomel for *Life*. 60, 61: Francis Miller for *Life*. 62, 63: NASA; Ralph Morse for *Life* (2). 64, 65: NASA. 66: N. R. Farbman, courtesy NASA—NASA. 67: NASA. 68, 69: Wide World. 70, 71: George Tames for *The New York Times*; John Dominis for *Life*. 72, 73: Ralph Morse for *Life*. 75: NASA. 77: TASS (2); Novosti from Opera Mundi, Milan. 78-83: Farrell Grehan for *Life*. 84, 85: Donald C. Uhrbrock for *Life*. 86: Donald C. Uhrbrock for *Life*—Jerry Cooke for *Time*. 88, 89: Ralph Morse for *Life*—NASA (2); Michael Rougier for *Life*. 90, 91: NASA—Donald C. Uhrbrock for *Life* (2). 92: NASA. 93: Henry Groskinsky for *Time*. 94, 95: NASA. 96, 97: NASA (2)—Robert Gomel for *Life*. 98, 99: NASA. 100, 101: NASA, except bottom right, Donald C. Uhrbrock for *Life*. 102, 103: NASA—Ralph Morse for *Life* (2). 104-113: NASA. 114: Ralph Morse for *Life*—NASA. 115: NASA. 116, 117: NASA, except upper right, Wide World. 118, 119: NASA. 120, 121: NASA, except upper left, Ralph Morse for *Life*. 123: Ralph Morse for *Life*. 125, 127: NASA. 128: Arthur Schatz for *Life*. 129: Yale Joel for *Life*—Arthur Schatz for *Life*. 130, 131: Yale Joel for *Life* (3); drawing by George Bell. 132, 133: CBS News Photos (3)—drawing by George Bell. 134, 135: Ralph Morse for *Life*. 136: Fritz Goro for *Life*—NASA—Ralph Morse for *Life*. 138: NASA—Ralph Morse for *Life*—NASA. 139: NASA. 140: David Scherman for *Life*. 141: Paul Schutzer for *Life*—Francis Miller for *Life*. 142: NASA. 143: NASA (2)—Vernon Merritt for *Life*. 144-147: NASA. 148, 149: NASA—Ralph Morse for *Life* (3); Vernon Merritt for *Life*. 150, 151: NASA, except bottom left, Francis Miller for *Life*. 152-157: NASA. 158, 159: NASA—Lynn Pelham for *Life*. 160, 161: Lynn Pelham for *Life*. 162-168: NASA. 169: NASA; Lynn Pelham for *Life*—NASA. 170, 171: NASA—Bill Eppridge for *Life*. 172, 173: Ralph Morse for *Life* (2); John Olson for *Life* (2)—NASA. 174, 175: NASA. 176, 177: Lynn Pelham for *Life*; NASA. 178, 179: NASA, except center left, UPI. 180, 181: NASA (3); Henry Groskinsky for *Life*—Fritz Goro for *Life*. 182: NASA—George Silk for *Life*—Ralph Morse for *Life*. 183: Ralph Morse for *Life*—NASA. 184: Ralph Morse for *Life*. 185-189: NASA. 190, 191: NASA—courtesy Time Inc. Picture Collection; NASA. 192-195: NASA. 197: Don Wright from *The Miami News*. 198, 199: Drawing by William J. Hennessey Jr.; NASA; Ralph Morse for *Time* (2). 200, 201: Ralph Morse for *Time*. 202-205: NASA. 206, 207: NASA (2)—drawings by Fred Holz. 208-215: NASA. 216, 217: NASA (2); Naval Research Laboratory. 218, 219: NASA. 220, 221: NASA (2)—Leonard Wright, Albany, Western Australia. 222, 223: NASA. 224, 225: From *The National Air and Space Museum* by C.D.B. Bryan with photographs by Michael Freeman, Robert Golden and Dennis Rolfe; published by Harry N. Abrams, Inc.—NASA (2). 226, 227: Ken Sherman for *Discover*. 229: USAF/Arnold Engineering Development Center—James L. Long and Associates © 1980; NASA. 230-233: Drawings by Fred Holz. 234: NASA/David Baker, Lincolnshire, England—NASA. 236, 237: NASA; Charlie Trainor from *The Miami News*. 238, 239: Joel Stevenson for *Life*; NASA (2). 240, 241: NASA, except bottom left, Wide World. 242: NASA—Carmine Ercolano for *Time*; Huber-The Orlando Sentinel/Skyline. 243: NASA. 244, 245: NASA—Hank Morgan for *Discover*. 246: NASA—Lief Ericksenn. 247-249: NASA. 250, 251: David Hume Kennerly for *Time*; NASA. 252-255: NASA. 256: Tony Suarez for *Time*. 257: NASA—Tony Suarez for *Time*. 258-263: NASA. 264, 265: NASA/Jet Propulsion Laboratory. 266: Drawing by William J. Hennessey Jr. 267: NASA. 268, 269: Frederic F. Bigio from B-C Graphics. 270: Frederic F. Bigio from B-C Graphics—TASS from Sovfoto. 271: NASA—Gene Gurney Collection (2). 272, 273: Ralph Crane for *Life*. 274: Michael Dressler for *Time*—NASA/Jet Propulsion Laboratory. 276, 277: Drawing by William J. Hennessey Jr.; drawing by Frederic F. Bigio from B-C Graphics—NASA (2). 278: Drawing by William J. Hennessey Jr.—NASA/Jet Propulsion Laboratory. 279: NASA (2); drawing by Frederic F. Bigio from B-C Graphics. 280: Drawing by William J. Hennessey Jr.—NASA (2). 281: Drawing by Frederic F. Bigio from B-C Graphics—NASA. 282, 283: Drawing by William J. Hennessey Jr.; drawing by Frederic F. Bigio from B-C Graphics—NASA. 284, 285: NASA/Jet Propulsion Laboratory (2); drawing by Frederic F. Bigio from B-C Graphics. 286, 287: NASA. 288: Drawing by William J. Hennessey Jr.—NASA/Ames Research Center. 289: Drawing by Frederic F. Bigio from B-C Graphics—NASA. 290, 291: Drawing by William J. Hennessey Jr.—drawing by Frederic F. Bigio from B-C Graphics; NASA/Jet Propulsion Laboratory. 292-297: NASA. 298, 299: NASA/Jet Propulsion Laboratory.

ACKNOWLEDGMENTS

The index for this book was prepared by Karla J. Knight. For their valuable help in the preparation of this book, the editors wish to thank: In California: Juri van der Woude, Jet Propulsion Laboratory, Pasadena. In Texas: Mike Gentry, NASA/Lyndon B. Johnson Space Center, Houston. In Virginia: Dr. John C. Houbolt, NASA/Langley Research Center, Hampton. In Washington, D.C.: Les Gaver, J. C. Hood, Althea Washington and Anita Williams, Audio-Visual Branch, National Aeronautics and Space Administration; Walter J. Boyne, Director, National Air and Space Museum, Smithsonian Institution.

BIBLIOGRAPHY

Allaway, Howard, *The Space Shuttle at Work*. National Aeronautics and Space Administration, 1979.

Armstrong, Neil, Michael Collins and Edwin E. Aldrin Jr. (with Gene Farmer and Dora Jane Hamblin), *First on the Moon: A Voyage with Neil Armstrong, Michael Collins, Edwin E. Aldrin, Jr.* Little, Brown and Company, 1970.

Beatty, J. Kelly, Brian O'Leary, Andrew Chaikin, eds., *The New Solar System*. Sky Publishing Corporation, 1981.

Belew, Leland F., ed., *Skylab: Our First Space Station*. National Aeronautics and Space Administration, 1977.

Benson, Charles D., and William Barnaby Faherty, *Moonport: A History of Apollo Launch Facilities and Operations*. National Aeronautics and Space Administration, 1978.

Bergaust, Erik, *Wernher von Braun*. National Space Institute, 1976.

Brooks, Courtney G., *Chariots for Apollo*. National Aeronautics and Space Administration, 1979.

Carpenter, M. Scott, et al., *We Seven*. Simon and Schuster, 1962.

Chappell, Russell E., *Apollo*. National Aeronautics and Space Administration, 1972.

Cooper, Henry S. F., Jr., *A House in Space*. Holt, Rinehart and Winston, 1976.

Corliss, William R., *The Interplanetary Pioneers*, Vol. 1, *Summary*. National Aeronautics and Space Administration, 1972.

Cortright, Edgar M., ed., *Apollo Expeditions to the Moon*. National Aeronautics and Space Administration, 1975.

Cromie, William J., *Skylab*. David McKay Company, Inc., 1976.

The Editors of Time-Life Records, *To the Moon*. Time-Life Records, 1969.

Fimmel, Richard O., James Van Allen and Eric Burgess, *Pioneer: First to Jupiter, Saturn, and Beyond*. National Aeronautics and Space Administration, 1980.

French, Bevan M., and Stephen P. Maran, eds., *A Meeting With the Universe: Science Discoveries from the Space Program*. National Aeronautics and Space Administration, 1981.

Gatland, Kenneth:
The Illustrated Encyclopedia of Space Technology. Harmony Books, 1981.
Missiles and Rockets. Macmillan Publishing Company, 1975.
Robot Explorers. The Macmillan Company, 1972.

Grissom, Betty, and Henry Still, *Starfall*. Thomas Y. Crowell Company, 1974.

Grissom, Virgil (Gus), *Gemini: A Personal Account of Man's Venture into Space*. The Macmillan Company, 1968.

Gurney, Gene, and Clare Gurney, *Cosmonauts in Orbit: The Story of the Soviet Manned Space Program*. Franklin Watts, 1974.

Hacket, Barton C., and James M. Grimwood, *On the Shoulders of Titans*. National Aeronautics and Space Administration, 1977.

Hallion, Richard P., and Tom D. Crouch, eds., *Apollo: Ten Years since Tranquility Base*. National Air and Space Museum, 1979.

Joels, Kerry Mark, and Gregory P. Kennedy. *The Space Shuttle Operator's Manual*. Ballantine Books, 1982.

Koenig, L. R., F. W. Murray, C. M. Michaux and H. A. Hyatt, *Handbook of the Physical Properties of the Planet Venus*. National Aeronautics and Space Administration, 1967.

Koppes, Clayton R., *JPL and the American Space Program: A History of the Jet Propulsion Laboratory*. Yale University Press, 1982.

Lehman, Milton, *This High Man: The Life of Robert H. Goddard*. Farrar, Straus and Company, 1963.

Michaux, C. M., et al., *Handbook of the Physical Properties of the Planet Jupiter*. National Aeronautics and Space Administration, 1967.

Morrison, David, *Voyages to Saturn*. National Aeronautics and Space Administration, 1982.

National Aeronautics and Space Administration, *In This Decade: Mission to the Moon*. 1971.

National Air and Space Museum, *Apollo to the Moon: A Dream of Centuries*. Smithsonian Institution, 1982.

Rosenthal, Alfred, *Satellite Handbook: A Record of NASA Space Missions 1958-1980*. NASA, Goddard Space Flight Center, 1981.

Sagan, Carl, *Cosmos*. Random House, 1980.

Skylab Experiments. National Aeronautics and Space Administration, 1972.

Stockton, William, and John Noble Wilford, *Spaceliner: The New York Times Report on the Columbia's Voyage into Tomorrow*. Times Books, 1981.

Swenson, Loyd S., Jr., James M. Grimwood and Charles C. Alexander, *This New Ocean: A History of Project Mercury*. National Aeronautics and Space Administration, 1966.

Tregaskis, Richard, *X-15 Diary*. E. P. Dutton & Company, 1961.

TRW Space Log: Twenty-fifth Anniversary of Space Exploration, 1957-1982. 1983.

Turnill, Reginald, *The Observer's Book of Manned Spaceflight*. London: Frederick Warne, 1978.

Washburn, Mark, *Distant Encounters: The Exploration of Jupiter and Saturn*. Harcourt Brace Jovanovich, 1983.

Wilding-White, T. M., *Jane's Pocket Book of Space Exploration*. Collier Books, 1976.

Wolfe, Tom, *The Right Stuff*. Farrar, Straus and Giroux, 1979.